VISUAL DICTIONARY
ANIMALS

VISUAL
DICTIONARY
ANIMALS

BARNES & NOBLE BOOKS
NEW YORK

This edition published by Barnes & Noble, Inc.
by arrangement with Fog City Press

2004 Barnes & Noble Books

Copyright © 2004 Weldon Owen Pty Ltd

All rights reserved. Unauthorized reproduction,
in any manner, is prohibited.
A catalog record for this book is available from
the Library of Congress, Washington, D.C.

M 10 9 8 7 6 5 4 3 2 1

ISBN 0-7607-6006-3

Color reproduction by Colourscan Co Pte Ltd
Printed by Tien Wah Press Pte Ltd
Printed in Singapore

Fog City Press
Chief Executive Officer: John Owen
President: Terry Newell
Publisher: Lynn Humphries
Creative Director: Sue Burk
Project Editor: Jessica Cox
Project Designer: Heather Menzies
Editorial Coordinator: Jennifer Losco
Consultants: George McKay, Richard Vogt,
Chris MacDonald, Hugh Dingle
Production Manager: Caroline Webber
Production Coordinator: James Blackman
Sales Manager: Emily Jahn
Vice President International Sales: Stuart Laurence

Produced using arkiva retrieval technology
For further information, contact arkiva@weldonowen.com.au

Contents

ANIMAL HABITATS — 8

PREHISTORIC LIFE — 32

MAMMALS — 74

BIRDS — 232

REPTILES 338

AMPHIBIANS 396

FISHES 422

INVERTEBRATES 486

FACT FILE 552

ANIMAL HABITATS

LIVING WORLD
Living World   12

CHANGING HABITATS
Changing Habitats   14

TROPICAL RAINFORESTS
Tropical Rainforests   16

SAVANNAS
Savannas   18

DESERTS
Deserts   20

TEMPERATE REGIONS
Temperate Regions   22

WOODLANDS
Woodlands   24

ALPINE REGIONS
Alpine Regions   26

SEASHORES
Seashores   28

OCEANS
Oceans   30

Living World

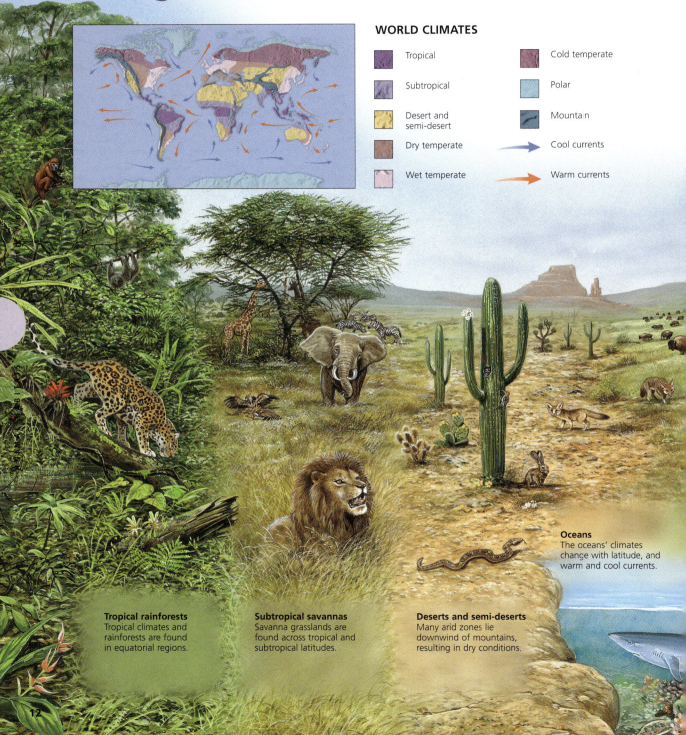

WORLD CLIMATES

- Tropical
- Subtropical
- Desert and semi-desert
- Dry temperate
- Wet temperate
- Cold temperate
- Polar
- Mountain
- Cool currents
- Warm currents

Oceans
The oceans' climates change with latitude, and warm and cool currents.

Tropical rainforests
Tropical climates and rainforests are found in equatorial regions.

Subtropical savannas
Savanna grasslands are found across tropical and subtropical latitudes.

Deserts and semi-deserts
Many arid zones lie downwind of mountains, resulting in dry conditions.

DIVERSE ECOSYSTEMS

Planet Earth's weather systems have created a range of habitats. Within each habitat is a community of plants and animals, called an ecosystem. To survive, individuals must adapt to their environment, making the most of the resources it has to offer.

Alpine and polar regions
These regions are typified by their harsh climates and barren landscapes.

Temperate grasslands
Dry temperate climates are found in mid-latitudes, and produce grasslands.

Temperate forests
Wet temperate climates are found in mid-latitudes, and produce forests.

Coniferous forests
Northern temperate (or boreal) climates occur in the northern hemisphere.

Changing Habitats

Carboniferous period
360–286 million years ago

Permian period
286–248 million years ago

Triassic period
248–208 million years ago

EVOLUTION OF HABITATS

Plants began to grow on Earth about 550 million years ago, creating the first environments for animal life. At first spore-bearing mosses and ferns dominated. Then seed-bearing conifers and flowering plants emerged. As plants evolved so did the animals that ate them, resulting in the complex ecosystems of today.

Carboniferous
Horsetail fern

Permian
Ginko

Triassic
Cyclad

Tropical Rainforests

BROMELIAD LIFE

Velvet worms can be found here, high above their usual forest-floor home.

Mosquitoes breed in the bromeliad pool.

Female poison frog deposits tadpoles into water-filled bromeliads.

Alligator lizard drinks from the pool, safe above the forest floor.

Beetle larvae prey on poison frog tadpoles.

Land snails favor the moist leaves for foraging and hiding.

Mouse opossum eats a damselfly larva from the bromeliad pool.

RAINFOREST PLANTS

Leaf dripping rainwater

Flowering ginger

Epiphytic plant

TROPICAL CLIMATE Hot, humid weather in the tropics provides the perfect conditions for lush rainforests, home to more animal species than any other environment.

Savannas

SAVANNA PLANTS

Sturt's pea

Wheat

Eucalyptus flower

GELADA BABOON HABITAT

Colored chest markings alter with a female's state of fertility.

Gelada baboons live in large social groups on the high grasslands of Ethiopia and Eritrea.

They are the only species of primate to eat grasses, which are supplemented by roots, stems and fruits.

SUBTROPICAL CLIMATE
Subtropical areas generally receive rainfall only in summer. This provides perfect conditions for savanna grasslands rather than dense forests.

Deserts

DESERT BY DAY
Plant and animal species that live in deserts have adapted to the lack of water. Some plants store water in their stems; others have long roots reaching underground.

DESERT LIFE

Jeweled gecko feeds on invertebrates in the spinifex, emitting a toxic fluid when startled.

Crimson chats live in open plains with a cover of spinifex or saltbush.

Small nocturnal cockroaches feed on decaying organic matter.

Ningauis are tiny marsupials that quench their thirst by licking dew from spinifex.

Burton's legless lizard preys on smaller lizards, suffocating and then swallowing them.

Desert skink eats termites and other invertebrates in the spinifex community.

DESERT BY NIGHT
Because of the intense heat of the day in some regions, many desert animals are nocturnal. Some plants have adapted to flower at night so they will be pollinated.

Beaver-tail cactus

Quiver tree

Century plant flowers

Chain-fruit cholla

Temperate Regions

TEMPERATE CLIMATE
Mid-latitude areas are typified by four distinct seasons. Depending on rainfall, temperate regions can be grasslands, shrublands or woodlands. To survive cold winters, animals may migrate or hibernate.

FOUR SEASONS

Spring: leaves sprout; flowers bloom.

Summer: fertilized flowers form fruits.

Autumn: fruits seed; leaves fall.

Winter: buds protect new growth.

TEMPERATE LIFE

Pika

Cuckoo

Cattle

Woodlands

WOODLAND PLANTS

Autumn leaves

Pine cones

Maple seeds

CONIFEROUS FORESTS
Evergreen conifers thrive in the forests of cold temperate regions. The trees are cone-shaped, allowing snow to slide off. Many animals have thick fur for warmth, while some hibernate to escape the bitter cold.

Gamebirds such as grouse have thick, downy plumage to survive the cold winters.

With coloring like mottled bark, owls are well suited to coniferous forests.

Wolverines are skilled hunters in deep snow. Their fur is hollow for insulation.

Solitary mooses, the largest of all deer, gather in groups only in the breeding season.

To survive winter, squirrels gather a cache of nuts and berries every autumn.

Alpine Regions

ALPINE PLANTS

Edelweiss

Puya

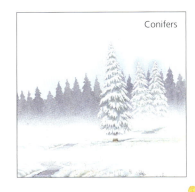
Conifers

HYRAX HABITAT

Rock and bush hyraxes live in close proximity.

Rock hyraxes feed on alpine grasses.

Bush hyraxes forage for leaves and bark.

ALPINE CLIMATE
As altitudes and latitudes increase, vegetation decreases, until trees can no longer grow. Mountain areas are similar to arctic tundra: plants and animals adapt to flourish in short summers and survive long winters.

Seashores

COASTAL PLANTS

Coconut sprouting

Screw pine fruit

Casuarina leaves

SEASHORES

Only salt-resistant plants can grow close to the sea, with grasses and shrubs forming the frontline against sea breezes. Below the splash line, tidal plants and animals thrive in the pools and among rocks.

Hermit crabs protect their soft bodies in discarded mollusk shells.

Mussels grow in dense colonies attached to rocks.

Clinging to rocks like mollusks, barnacles are actually crustaceans.

Blennies hide under stones, darting from crevice to shadow.

Carnivorous starfish move slowly on hundreds of feet.

Small octopuses are secretive inhabitants of tidepools.

Seaweeds are marine algae, and lack a plant's roots, flowers, seeds and fruit.

Colorful sea anemones feed on small animals stung by their tentacles.

Sculpin are predators that lie in wait for prey rather than chase it.

Crabs are abundant in intertidal tidepools.

Oceans

OCEAN HABITATS
The oceans contain many habitats, dependent on latitude and depth. A huge variety of animals and plants have adapted to life underwater, from seaweeds and seagrasses, to sharks, corals and fishes.

Dugongs live among the coastal seagrass beds in the Indo-Pacific region.

Yellow-finned leatherjackets are herbivorous fishes.

Blue swimmer-crabs bury themselves in sand up to their eyes to wait for prey.

Fan-mussels anchor into the sand with only part of their shell exposed.

OCEAN LIFE

Solitary green turtles migrate vast distances to lay their eggs.

Sunlight zone
Enough sunlight penetrates this layer for plants to live. It contains most of the ocean's life.

Twilight zone
Not enough sunlight to sustain plant life. Many fishes in this zone contain light-producing bacteria and glow in the dark.

Midnight zone
This is a vast mass of cold, slow-moving water in absolute darkness. Life is about a tenth that of the twilight zone and is also less diverse.

Abyssal zone
The near-freezing, pitch-black bottom layer has few species and little food. Near hydrothermal vents, some animals make food using internal bacteria and hydrogen sulfide.

0
650 ft (200 m)
3,250 ft (1,000 m)
9,850 ft (3,000 m)

PREHISTORIC LIFE

PREHISTORIC ERAS

Web of Life	36
Before the Dinosaurs	38
Triassic Period	40
Jurassic Period	42
Cretaceous Period	44

INTRODUCING DINOSAURS

Types of Dinosaurs	46
Dinosaur Characteristics	48
Dinosaur Hips	50
Raising Young	52

DINOSAUR SPECIES

Sauropods	54
Theropods	56
Ornithopods	60
Armored Dinosaurs	62
Ceratopians	64
Dinosaur Contemporaries	66

AFTER THE DINOSAURS

Dinosaur Extinction	68
Recreating the Dinosaurs	70
Modern Relatives	72

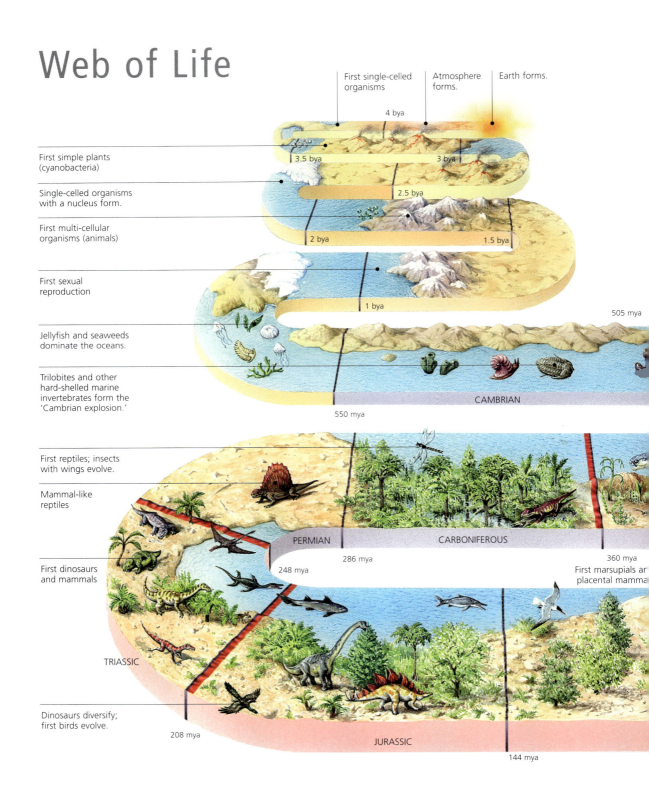

FOSSIL RECORD

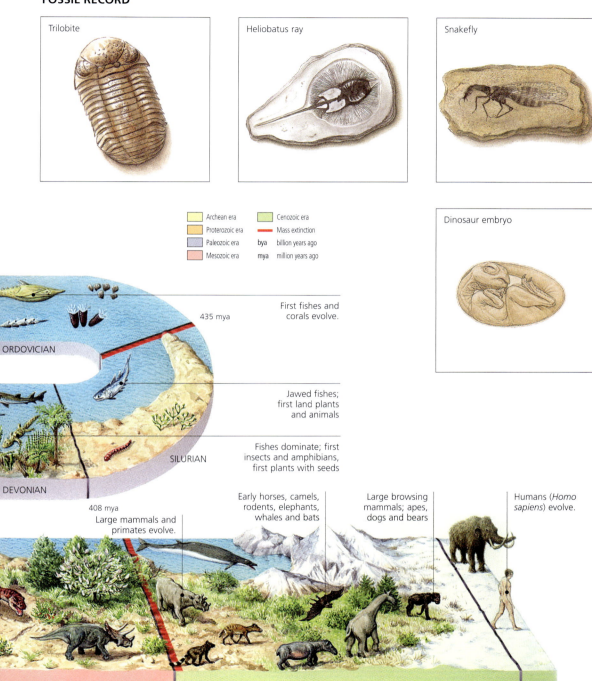

Before the Dinosaurs

Ornithosuchus | Dimetrodon | Dragonfly | Hylonomus | Ichthyostega | Dunkleosteus

38

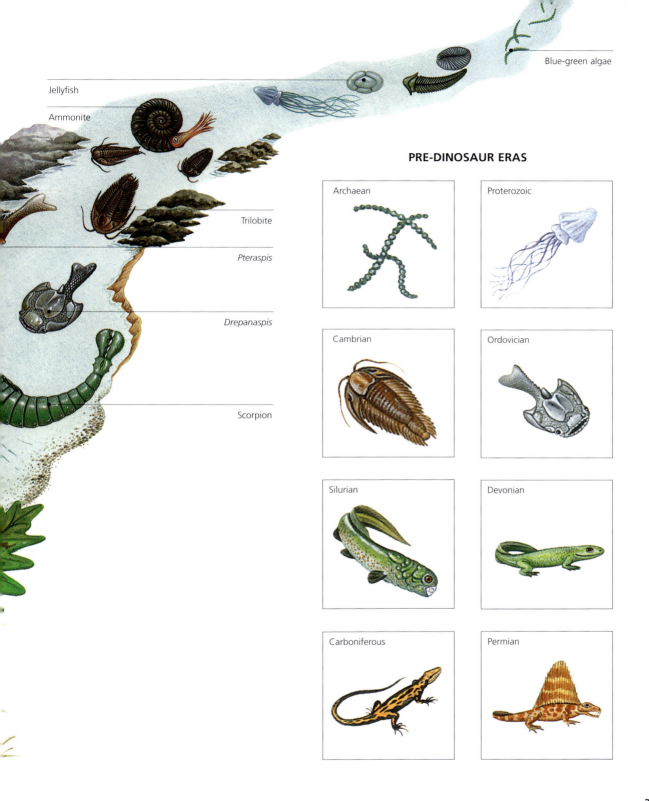

Triassic Period

Triassic landmasses

TRIASSIC LANDSCAPE

TRIASSIC DINOSAUR DIETS

Dragonfly

Wielandiella

Haramiya

Tree fern

Jurassic Period

Jurassic landmasses

JURASSIC LANDSCAPE

JURASSIC DINOSAUR DIETS

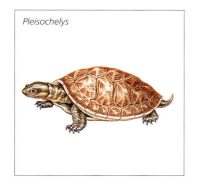

Pleisochelys

Archaeopteryx

Cockroach

Cretaceous Period

Cretaceous landmasses

CRETACEOUS LANDSCAPE

ORNITHOMIMUS

VELOCIRAPTOR

CHANGING DINOSAURS
Due to the continents moving apart and local conditions changing, these two duckbilled dinosaurs evolved differently from a similar ancestor.

SALTASAURUS

TYRANNOSAURUS

KRONOSAURUS

TRICERATOPS

CORYTHOSAURUS   PACHYCEPHALOSAURUS   EUOPLOCEPHALUS

CRETACEOUS LIFE

Polyglyphanodon

Magnolia

Crusafontia

Types of Dinosaurs

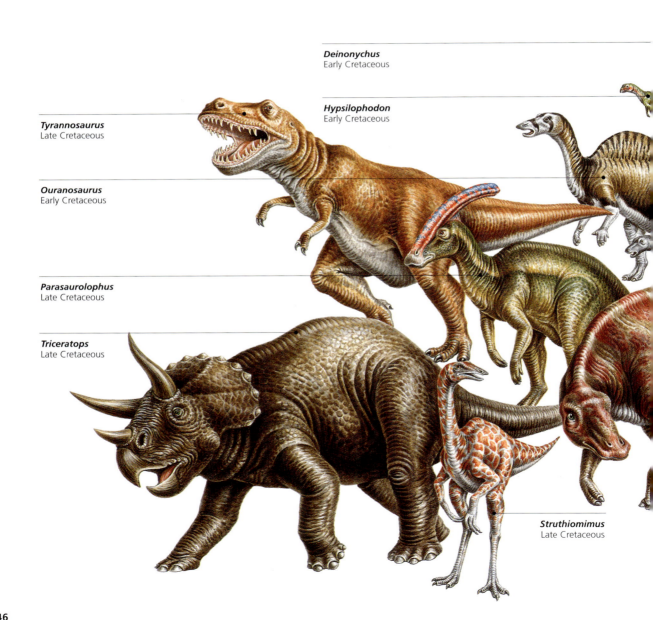

Deinonychus
Early Cretaceous

Tyrannosaurus
Late Cretaceous

Hypsilophodon
Early Cretaceous

Ouranosaurus
Early Cretaceous

Parasaurolophus
Late Cretaceous

Triceratops
Late Cretaceous

Struthiomimus
Late Cretaceous

Dinosaur Characteristics

DINOSAUR ANATOMY

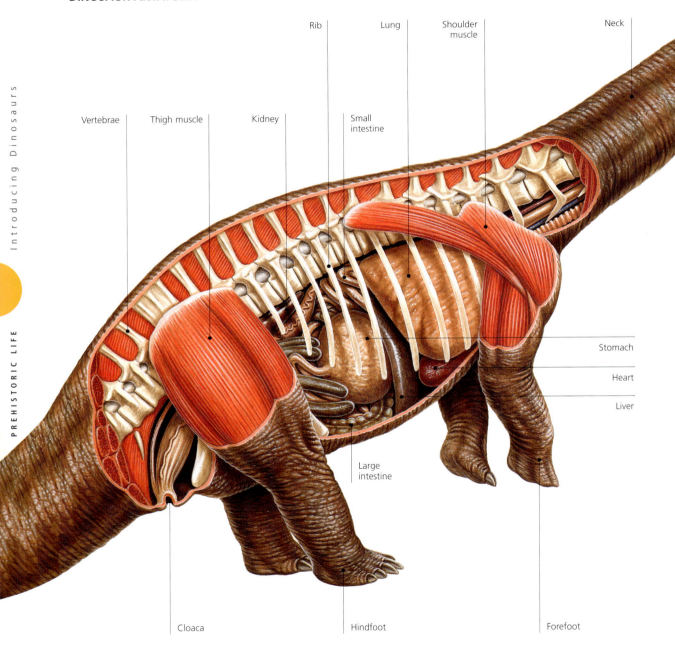

Head

BRAIN COMPARISON

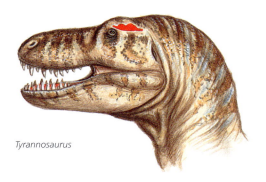
Tyrannosaurus

Troödon

Stegosaurus

Human

TYPES OF TEETH

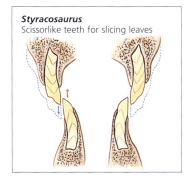
Styracosaurus
Scissorlike teeth for slicing leaves

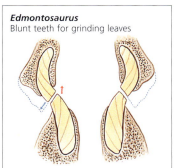
Edmontosaurus
Blunt teeth for grinding leaves

Dinosaur Hips

EVOLUTION OF HIPS

Legs sprawling on either side of body

Legs partially under body

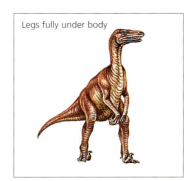
Legs fully under body

EDMONTOSAURUS
Bird-hipped dinosaur

Pubis points backward; allows space for large intestines of quadrapedal (four-legged) plant eaters.

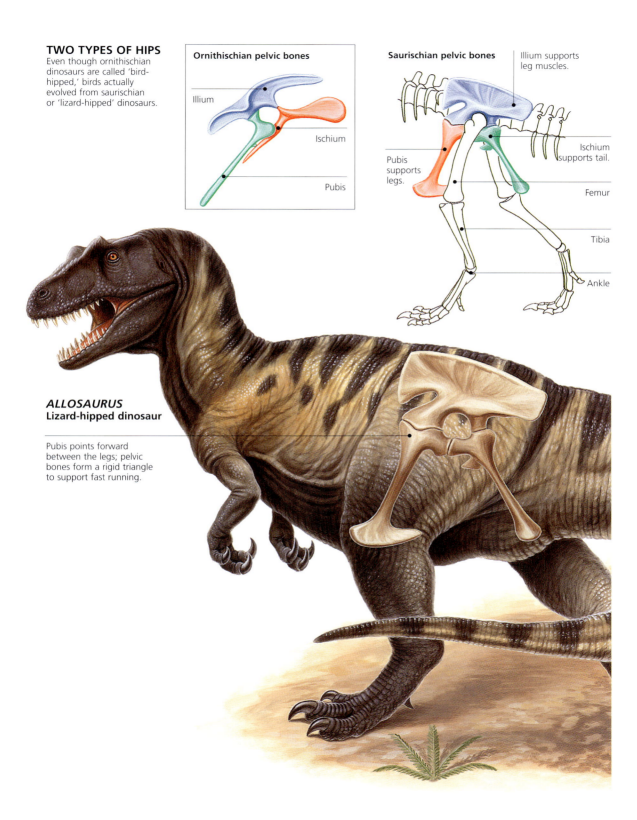

TWO TYPES OF HIPS

Even though ornithischian dinosaurs are called 'bird-hipped,' birds actually evolved from saurischian or 'lizard-hipped' dinosaurs.

Ornithischian pelvic bones
- Illium
- Ischium
- Pubis

Saurischian pelvic bones
- Illium supports leg muscles.
- Ischium supports tail.
- Pubis supports legs.
- Femur
- Tibia
- Ankle

ALLOSAURUS
Lizard-hipped dinosaur

Pubis points forward between the legs; pelvic bones form a rigid triangle to support fast running.

Raising Young

HATCHING OUT

Maiasaura hatchling

One-year-old *Maiasaura*

Adult *Maiasaura*

INSIDE AN EGG

Allantois stored waste.

Membrane (chorion) provided oxygen.

Albumen (yolk sac) nourished embryo.

Eggshell protected embryo.

Amniotic (fluid) sac cushioned dinosaur.

***OVIRAPTOR* NEST**
Mother *Oviraptor* feeds a freshly killed young *Velociraptor* to her nest of hatchlings. Unlike most modern birds and reptiles, some dinosaurs cared for their young after hatching.

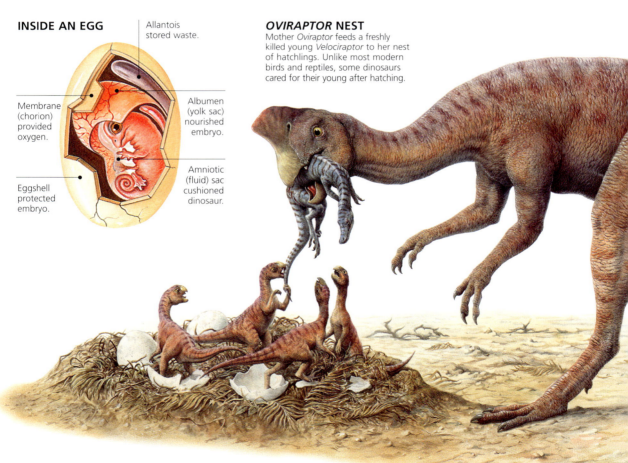

DINOSAUR NEST

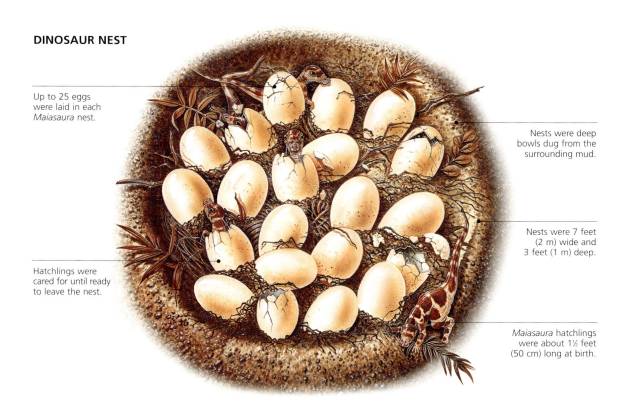

Up to 25 eggs were laid in each *Maiasaura* nest.

Nests were deep bowls dug from the surrounding mud.

Nests were 7 feet (2 m) wide and 3 feet (1 m) deep.

Hatchlings were cared for until ready to leave the nest.

Maiasaura hatchlings were about 1½ feet (50 cm) long at birth.

EGG ARRANGEMENTS

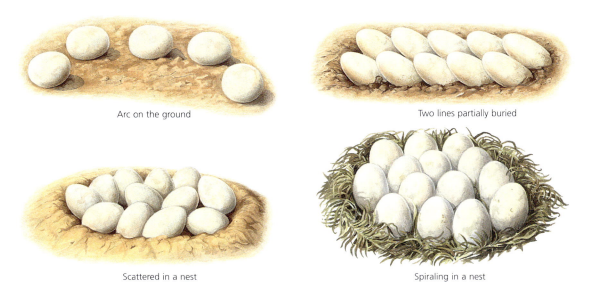

Arc on the ground

Two lines partially buried

Scattered in a nest

Spiraling in a nest

Sauropods

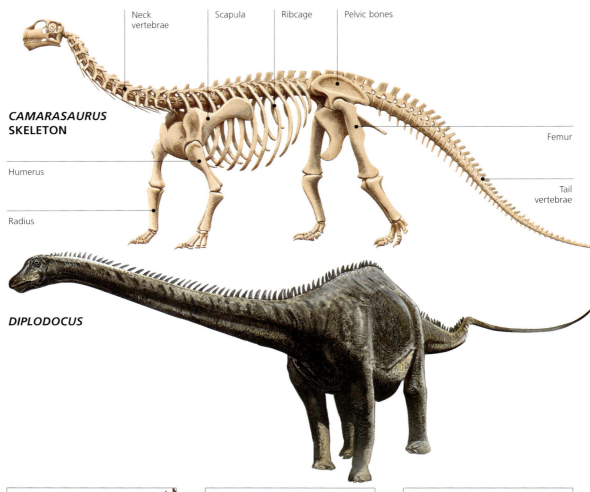

CAMARASAURUS SKELETON

Labels: Neck vertebrae, Scapula, Ribcage, Pelvic bones, Humerus, Radius, Femur, Tail vertebrae

DIPLODOCUS

Diplodocus tail

Diplodocus tail vertebra

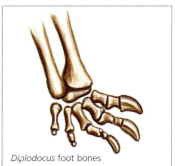

Diplodocus foot bones

PLATEOSAURUS

Plateosaurus skull

BRACHIOSAURUS

Brachiosaurus head

Brachiosaurus skull

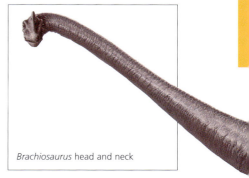

Brachiosaurus head and neck

Theropods

STRUTHIOMIMUS SKELETON

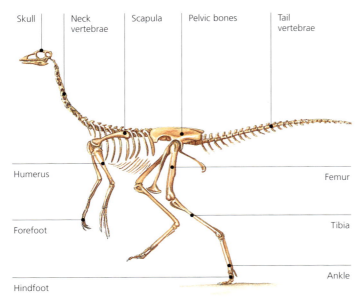

Skull · Neck vertebrae · Scapula · Pelvic bones · Tail vertebrae · Humerus · Forefoot · Hindfoot · Femur · Tibia · Ankle

EATING TOOLS

Allosaurus **skull**
Hinged jaws to swallow prey whole

Carcharodontosaurus **teeth**
Serrated teeth front and back to slice flesh

STRUTHIOMIMUS

ADAPTATIONS FOR SURVIVAL

Tyrannosaurus had hinged jaws and backward-facing teeth to rip and tear flesh.

Gallimimus ran at speeds of up to 30 miles (48 km) an hour. A slower dinosaur such as *Albertosaurus* had no chance of catching it.

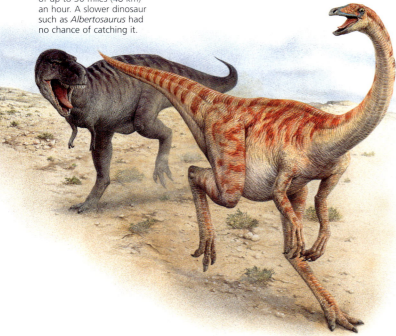

FEEDING

Baryonyx

Oviraptor

Albertosaurus

Compsognathus

Theropods

VELOCIRAPTOR

Velociraptor arm bones

Velociraptor foot

MEGALOSAURUS

Megalosaurus tooth

Serrated edge

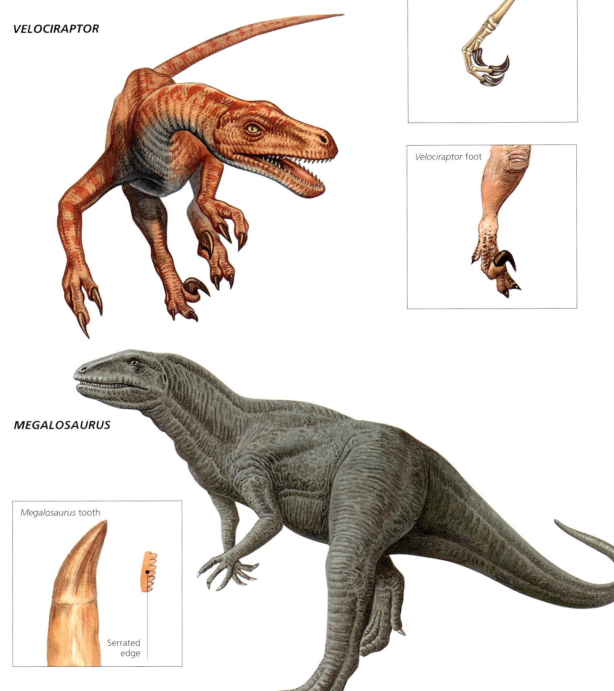

ALLOSAURUS

Allosaurus skull

OVIRAPTOR

Two Oviraptor heads showing crests

Ornithopods

OURANOSAURUS SKELETON

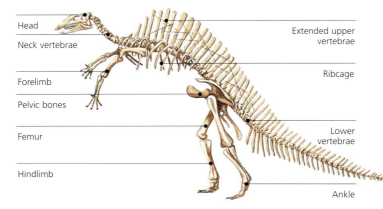

- Head
- Neck vertebrae
- Forelimb
- Pelvic bones
- Femur
- Hindlimb
- Extended upper vertebrae
- Ribcage
- Lower vertebrae
- Ankle

ORNITHOPOD SKULLS

Ouranosaurus skull

Anatotitan skull

Lambeosaurus skull

Corythosaurus skull

OURANOSAURUS

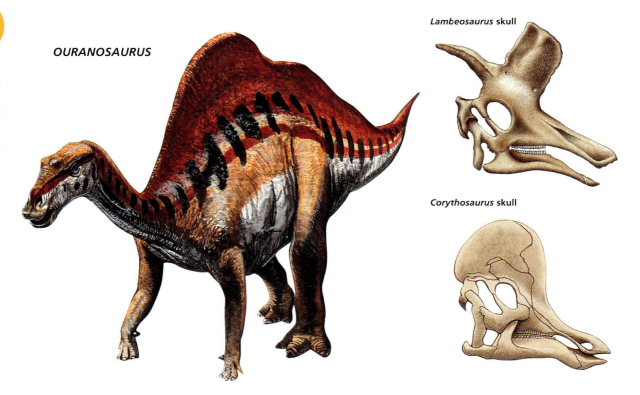

IGUANODONS

PARASAUROLOPHUS

Juvenile

Adult female

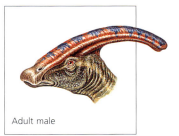
Adult male

Armored Dinosaurs

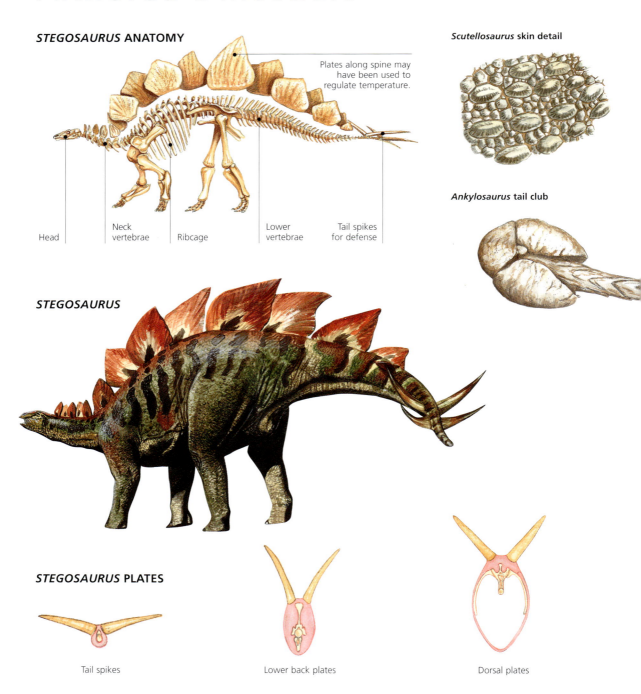

TUOJIANGOSAURUS

Tuojiangosaurus tail
Spikes used to stab potential predators.

EUOPLACEPHALUS

Euoplacephalus tail
Clublike bony tail used to hit predators.

Euoplacephalus skull
Armor-plated with spikes to protect neck

EDMONTONIA

Edmontonia head
Forward-facing spikes used to ram predators.

Ceratopians

CHASMOSAURUS HEAD

CHASMOSAURUS

STYRACOSAURUS SKULL

STYRACOSAURUS

TRICERATOPS

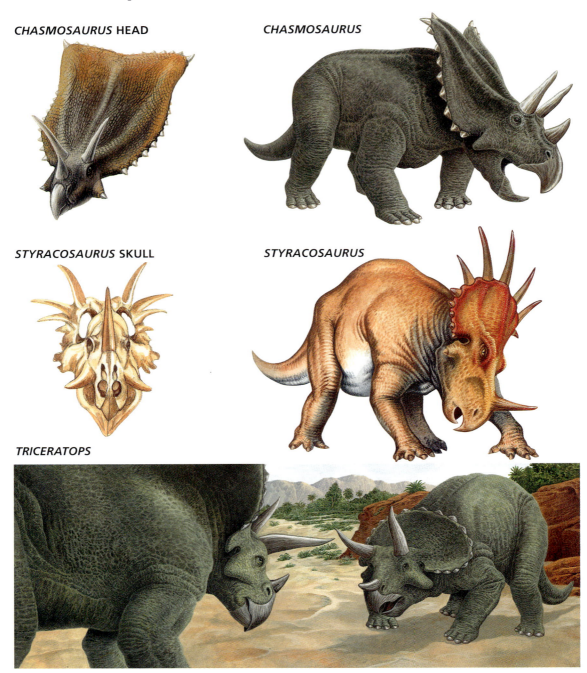

PROTOCERATOPS

PROTOCERATOPS SKULL

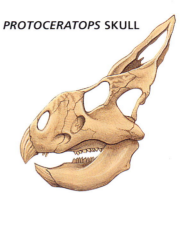

TOROSAURUS SKULL

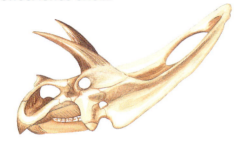

TOROSAURUS

CENTROSAURUS

CENTROSAURUS HEAD

Dinosaur Contemporaries

DIMORPHODON

PTERODAUSTRO

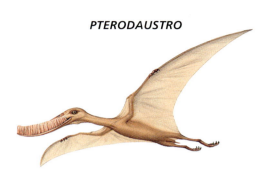

ARCHAEOPTERYX

DIMETRODON

MESONYX

ALPHADON

Dinosaur Extinction

EXTINCTION THEORIES

Hot climate change

Cold climate change

Volcanic eruptions

Meteorite crashing to Earth

DINOSAUR EXTINCTION
There are several theories about what caused the dinosaurs to suddenly disappear 65 million years ago. The most widely accepted theory is that a huge meteorite hit Earth causing environmental chaos.

Small mammal

AGE OF MAMMALS
The small mammals of the Cretaceous were some of the species that survived the mass extinctions of 65 million years ago. These evolved and diversified into the thousands of known species today.

Recreating the Dinosaurs

FOSSIL FORMATION

Safe from scavengers
A dead dinosaur lies decomposing at the bottom of a lake, its skeleton remaining intact.

Turning to stone
Trapped and flattened by layers of sedimentary rock, dinosaur bones are replaced by minerals.

Covered up
Layers of sand or silt cover the dinosaur's bones, which stops them from being washed away.

Fossil uncovered
Millions of years later, movements within the Earth bring the fossilized skeleton to the surface.

Pieces of fossilized tail vertebrae

Missing sections of vertebrae are sketched from surrounding bones.

Sketches are painstakingly executed.

WALKING ON THE PAST

DINOSAUR BONES

Tail vertebrae

Scapula

Hip bone

Femur

Neck vertebra

Rib

RECONSTRUCTING A DINOSAUR

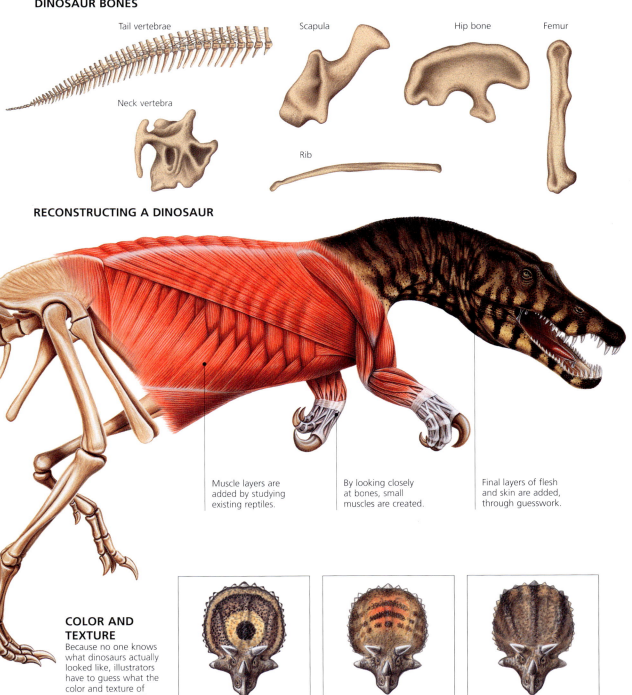

Muscle layers are added by studying existing reptiles.

By looking closely at bones, small muscles are created.

Final layers of flesh and skin are added, through guesswork.

COLOR AND TEXTURE
Because no one knows what dinosaurs actually looked like, illustrators have to guess what the color and texture of dinosaur skin was like.

Modern Relatives

ARCHAEOPTERYX
It is generally believed that modern birds evolved from dinosaurs. The skeleton of *Archaeopteryx*, the earliest known bird, is very similar to certain dinosaurs—such that many now classify *Archaeopteryx* as a dinosaur with feathers.

FROM DINOSAUR TO BIRD

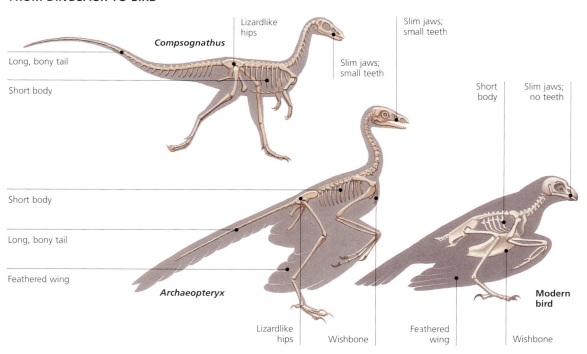

Compsognathus — Long, bony tail; Short body; Lizardlike hips; Slim jaws; small teeth

Archaeopteryx — Short body; Long, bony tail; Feathered wing; Lizardlike hips; Wishbone; Slim jaws; small teeth

Modern bird — Short body; Feathered wing; Wishbone; Slim jaws; no teeth

Waterfowl	Turtle	Crocodile
		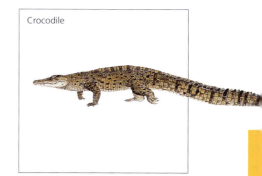
Fish	Emu	Seabird
	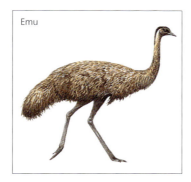	
Hoatzin	Lizard	Insect

Archaeopteryx wing	Bat wing	Pigeon wing
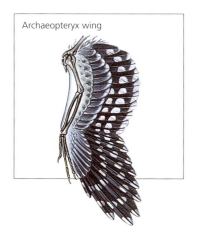		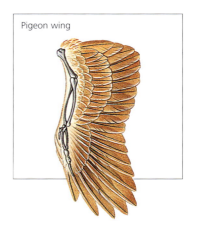

MAMMALS

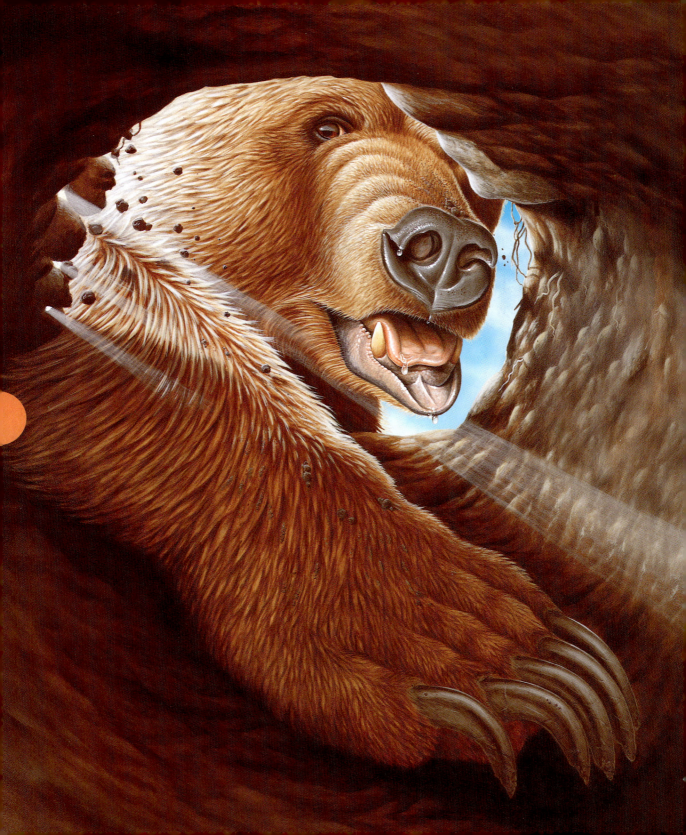

INTRODUCING MAMMALS

Classifying Mammals	78
Mammal Characteristics	80
Evolution of Mammals	82

MONOTREMES

Monotremes	84

MARSUPIALS

Marsupials	86
Kangaroos and Wallabies	90

XENARTHRANS AND PANGOLINS

Xenarthrans and Pangolins	92

INSECT-EATING MAMMALS

Insect-eating Mammals	94

FLYING LEMURS AND TREE SHREWS

Flying Lemurs and Tree Shrews	96

BATS

Bats	98
Old World Fruit Bats	100
Insect-eating Bats	102

PRIMATES

Primates	104
Lower Primates	106
Higher Primates	110
African Old World Monkeys	112
Asian Old World Monkeys	114
New World Monkeys	116
Great Apes	118
Gorillas	120

CARNIVORES

Carnivores	122
Wolves, Dogs and Foxes	124
Bears	130
Eurasian and New World Bears	132
Polar Bear	134
Asian Bears	136
Mustelids	138
Catlike Carnivores	142
Raccoons and Hyenas	144
Hyena Hunting Tactics	146
Cats	148
Great Cats	150
Lynxes and Cheetah	154
Small Cats	156
Seals, Sealions and Walrus	158

WHALES, DOLPHINS AND PORPOISES

Whales, Dolphins and Porpoises	162
Whale Anatomy	164
Baleen Whales	166
Toothed Whales	170
Dolphins	174
River Dolphins and Porpoises	176

HOOFED MAMMALS

Hoofed Mammals	178
Hippopotamuses, Pigs and Peccaries	180
Old World Camels	182
New World Camelids	184
Giraffe, Okapi and Pronghorn	186
Deer	188
Bovids	192
Antelopes	194
Wildebeest Migration	196
True Cattle, Goats and Sheep	198
Rhinoceroses	200
Horses, Asses and Zebras	202

AARDVARK, HYRAXES AND TAPIRS

Aardvark, Hyraxes and Tapirs	206

ELEPHANTS

Elephants	208

DUGONG AND MANATEES

Dugong and Manatees	212

ELEPHANT SHREWS

Elephant Shrews	214

RODENTS

Rodents	216
Squirrel-like Rodents	218
Beavers	222
Mouselike Rodents	224
Cavylike Rodents	226

RABBITS, HARES AND PIKAS

Rabbits, Hares and Pikas	228

Classifying Mammals

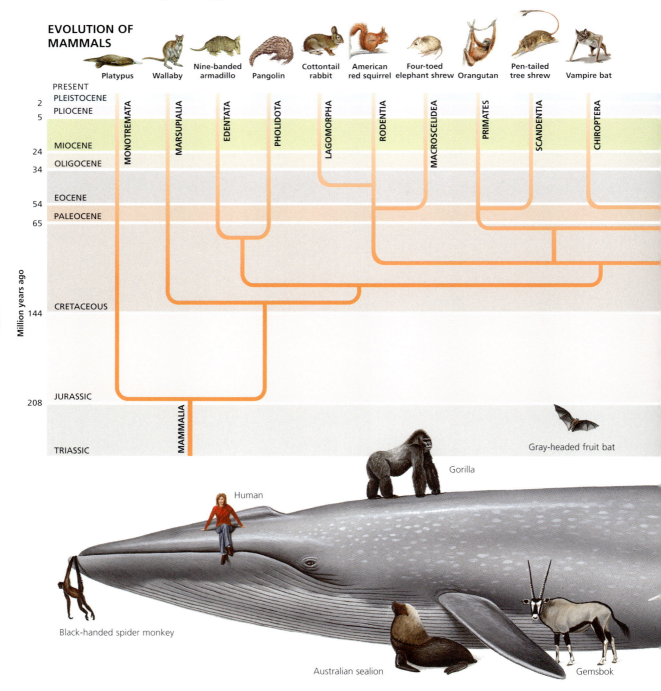

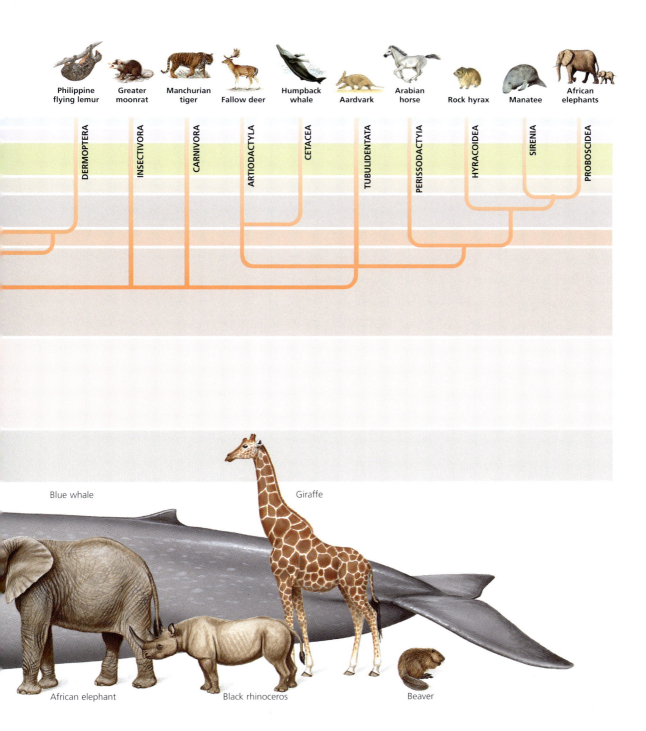

Mammal Characteristics

HUNTERS AND PREY

Forward-facing eyes for sight in three dimensions

Sideways-facing eyes for 180° vision

Heavily muscled hindlimbs for rapid acceleration

Claws to grasp prey

Slender hindlimbs for lightness with power

Toes modified to form hooves for greater leverage when running

MAMMAL FOREFEET

Whale: modified flipper for swimming

Bear: adapted for digging and hunting

Bat: fingers stretched for flying

GIVING BIRTH TO LIVE YOUNG
All mammals are vivaporous: they give birth to live young.

Wildebeest birth
Social mammals time mating so their young are born together, to protect them from predators.

After birth
Newly born wildebeests are able to stand and run within minutes of birth.

MAMMAL EAR
The three bones that make up a mammal's middle ear work together to transmit sound waves into the sensory cells of the inner ear.

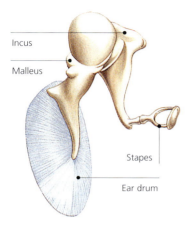

CONVERGENT EVOLUTION
Animal species that are not closely related often adapt to environmental conditions or pressures in similar ways. This is called convergent evolution.

MAMMAL HAIR
All mammals have hair on their bodies. A typical hair contains three layers: an insulating medulla, cortex containing pigment and outer cuticle.

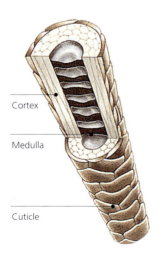

Evolution of Mammals

DIMETRODON
Mammal-like reptile that lived 300 million years ago.

MEGAZOSTRODON
Earliest known true mammal, it lived 220 million years ago.

CYNOGNATHUS
Mammal-like reptiles that lived 245–230 million years ago.

MORGANUCODONTID
One of the earliest true mammals, it lived 195 million years ago.

UINTATHERIUM
One of the earliest herbivores, it lived 60 million years ago.

ARSINOITHERIUM
Relative of modern hyraxes, it lived 40 million years ago.

INDRICOTHERIUM
The largest land mammal ever lived 30 million years ago.

URSAVUS
Thought to be the ancestor of modern bears, it lived 20 million years ago.

WOOLLY MAMMOTHS
Ice-age mammals related to modern elephants that lived 350–10 thousand years ago.

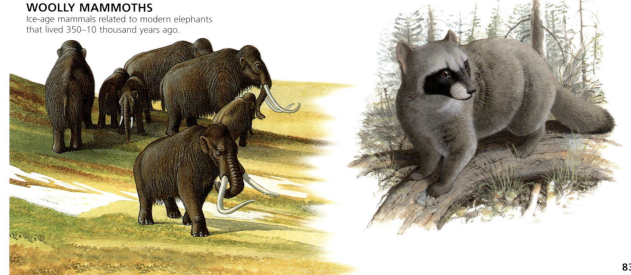

Monotremes

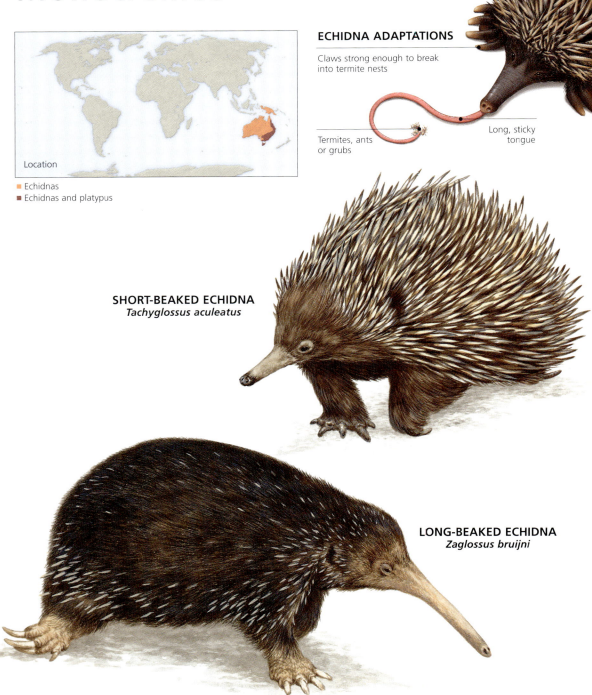

Location
- Echidnas
- Echidnas and platypus

ECHIDNA ADAPTATIONS
Claws strong enough to break into termite nests
Termites, ants or grubs
Long, sticky tongue

SHORT-BEAKED ECHIDNA
Tachyglossus aculeatus

LONG-BEAKED ECHIDNA
Zaglossus bruijni

WATERTIGHT EYES

Eyes open when out of water

Eyes closed when diving

PLATYPUS
Ornithorhynchus anatinus

PLATYPUS NEST
Monotremes are the only mammals that lay eggs and do not give birth to live young.

CLAWS FOR DIGGING

Web on forefoot drawn back

Web on forefoot extended

Marsupials

Location
- American marsupials
- Australasian marsupials

NUMBAT
Myrmecobius fasciatus

REPRODUCTIVE ORGANS

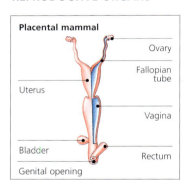

Placental mammal
- Ovary
- Fallopian tube
- Uterus
- Vagina
- Bladder
- Rectum
- Genital opening

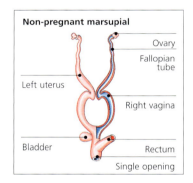

Non-pregnant marsupial
- Ovary
- Fallopian tube
- Left uterus
- Right vagina
- Bladder
- Rectum
- Single opening

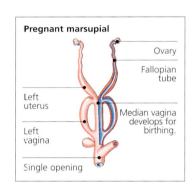

Pregnant marsupial
- Ovary
- Fallopian tube
- Left uterus
- Median vagina develops for birthing.
- Left vagina
- Single opening

STARTING OUT

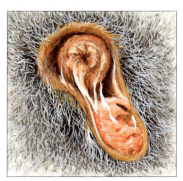
Embryonic young leaves birth canal.

It makes its way to its mother's pouch.

Finds nipple inside the pouch and suckles.

JUMPING INTO THE POUCH

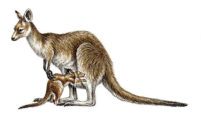

The young kangaroo pushes front feet and head into the pouch.

It rolls forward into the pouch.

Then twists around so it can see and jump out of the pouch.

RED KANGAROO
Macropus rufus

STRIPED POSSUM
Dactylopsila trivirgata

TASMANIAN DEVIL
Sarcophilus laniarius

Kangaroos and Wallabies

GOODFELLOW'S TREE KANGAROO
Dendrolagus goodfellowi

Strong forelimbs for climbing

FOREST WALLABY
Macropus sp.

MUSKY RAT KANGAROO
Hypsiprymnodon moschatus

WESTERN GRAY KANGAROO
Macropus fuliginosus

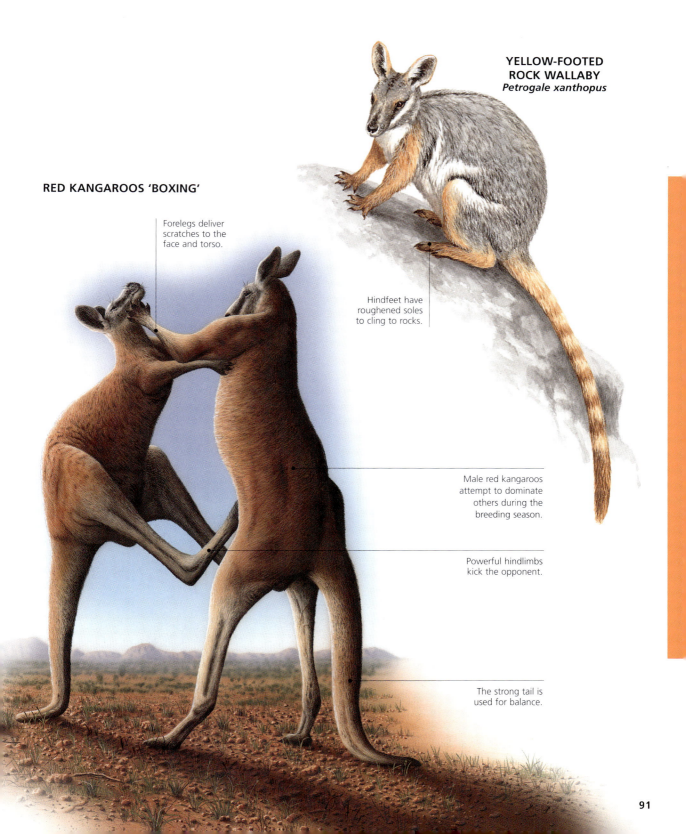

Xenarthrans and Pangolins

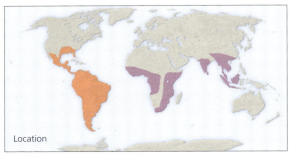

Location
- Xenarthrans (anteaters, armadillos and sloths)
- Pangolins

TAMANDUA
Tamandua sp.

MANED THREE-TOED SLOTH
Bradypus torquatus

THREE-BANDED ARMADILLO
Tolypeutes tricinctus

PANGOLIN
Phataginus sp.

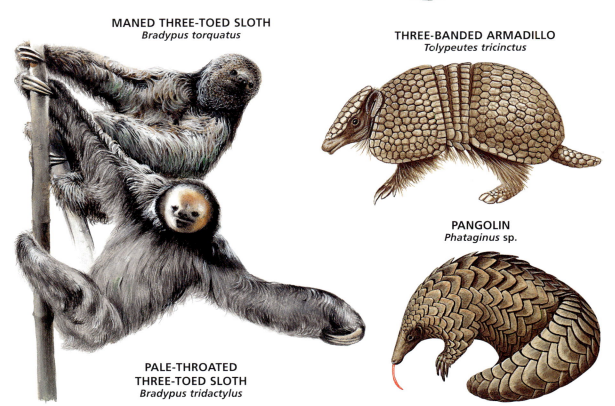

PALE-THROATED THREE-TOED SLOTH
Bradypus tridactylus

Insect-eating Mammals

Location
■ Insect-eating mammals

WESTERN EUROPEAN HEDGEHOG
Erinaceus europaeus

TYPES OF NOSES

Algerian hedgehog
Short pointed snout with bristles

European mole
Sensitive nose to smell and feel for prey

Pyrenean desman
Flexible probe to find insects underwater

MOONRAT
Echinosorex gymnura

PYRENEAN DESMAN
Galemys pyrenaicus

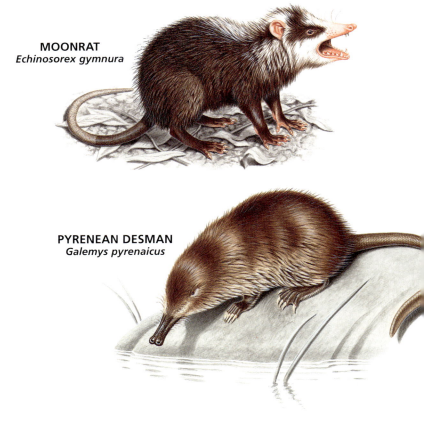

Flying Lemurs and Tree Shrews

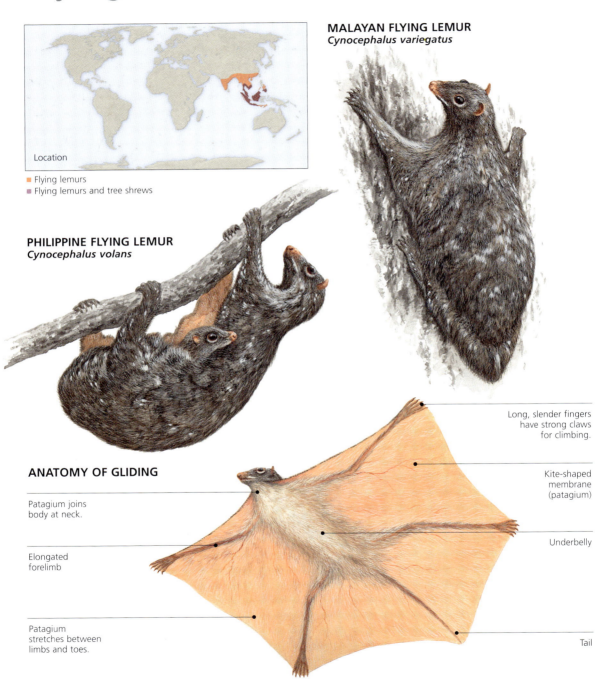

Location
- Flying lemurs
- Flying lemurs and tree shrews

MALAYAN FLYING LEMUR
Cynocephalus variegatus

PHILIPPINE FLYING LEMUR
Cynocephalus volans

ANATOMY OF GLIDING

Patagium joins body at neck.

Elongated forelimb

Patagium stretches between limbs and toes.

Long, slender fingers have strong claws for climbing.

Kite-shaped membrane (patagium)

Underbelly

Tail

Bats

BAT ANATOMY

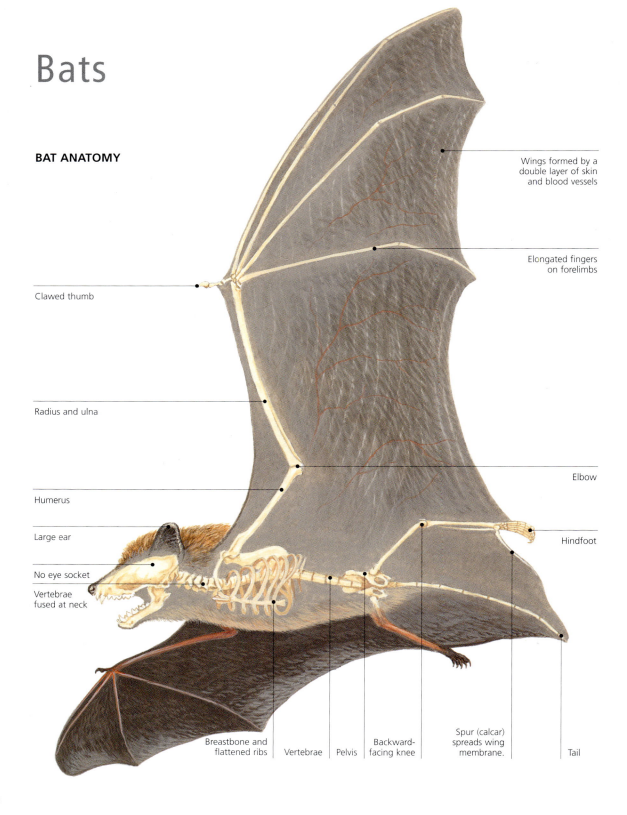

ECHOLOCATION
Insect-eating bats use pulses of ultrasound to detect and capture prey.

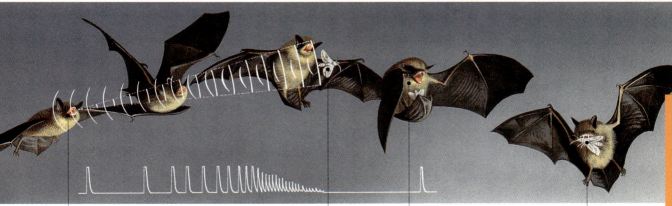

Bat emits rapid 'clicks' of sound.

The 'clicks' bounce off prey back to bat, revealing its location.

Frequency of 'clicks' increases as bat nears to pinpoint prey.

Bat seizes prey.

VARIATIONS

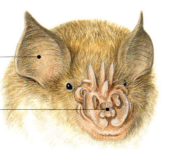

Trident bat

Ear detects reflected sound.

Nose emits 'clicks' of sound.

Whiskered bat

Ear detects reflected sound.

Mouth emits 'clicks' of sound.

EARS AND NOSES

Long-eared bat
Large ears with flaps may improve accuracy of echolocation.

Lesser bare-backed fruit bat
Tube-shaped nostrils sniff out food.

Tent-building bat
Flaps of skin on leaf-shaped nose help detect echoes from prey.

Old World Fruit Bats

Location
■ Old World fruit bats

GAMBIAN EPAULETTED FRUIT BAT
Epomophorus gambianus

HAMMER-HEADED FRUIT BAT
Hypsignathus monstrosus

GRAY-HEADED FLYING FOX
Pteropus poliocephalus

EGYPTIAN FRUIT BAT
Rousettus egyptiacus

FRUIT BAT EATING
Many fruit bats eat upside down, clutching their meal in their claws. Their relatively well-developed second digit and claw aids in this process.

INDIAN FLYING FOX
Pteropus giganteus

Insect-eating Bats

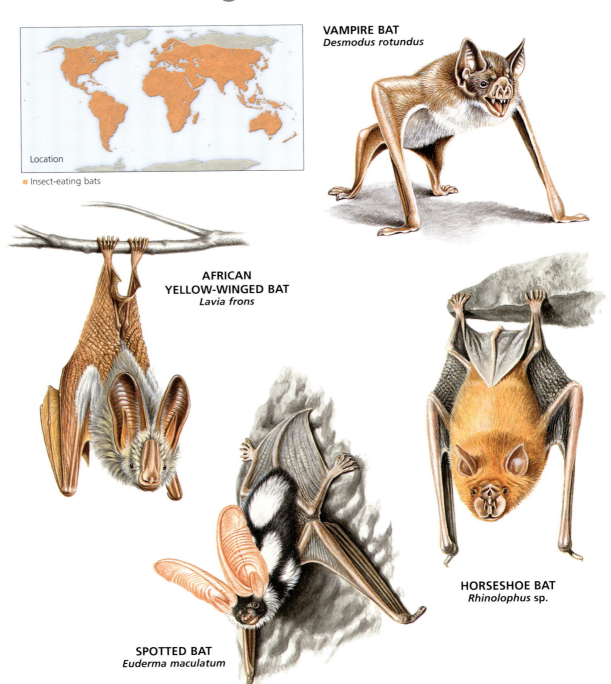

Location
- Insect-eating bats

VAMPIRE BAT
Desmodus rotundus

AFRICAN YELLOW-WINGED BAT
Lavia frons

SPOTTED BAT
Euderma maculatum

HORSESHOE BAT
Rhinolophus sp.

GREATER BULLDOG BAT
Noctilio leporinus

HONDURAN WHITE BATS
Ectophylla alba
Honduran white bats cut away the connection between the midrib and edge of palm fronds, causing the fronds to curl. They then roost in the 'tent' they have created.

WRINKLE-FACED BAT
Centurio senex

TENT-BUILDING BAT
Uroderma bilobatum

SWORD-NOSED BAT
Lonchorhina sp.

Primates

EVOLUTION OF PRIMATES

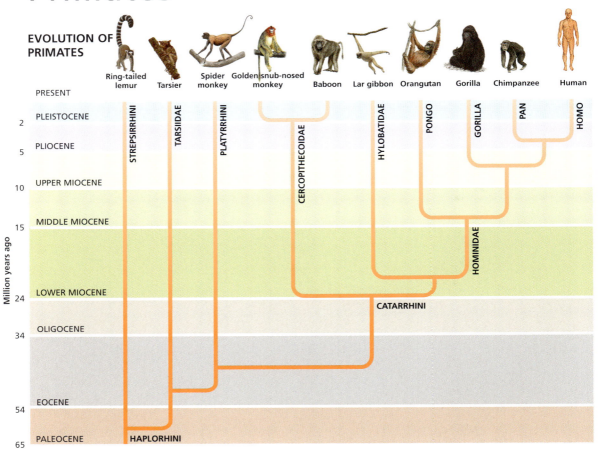

PRIMATE FEET AND HANDS

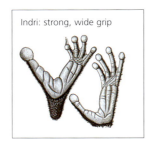

Indri: strong, wide grip

Aye-aye: uses claws to climb

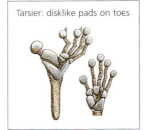

Tarsier: disklike pads on toes

Gorilla: broad to support weight

Lower Primates

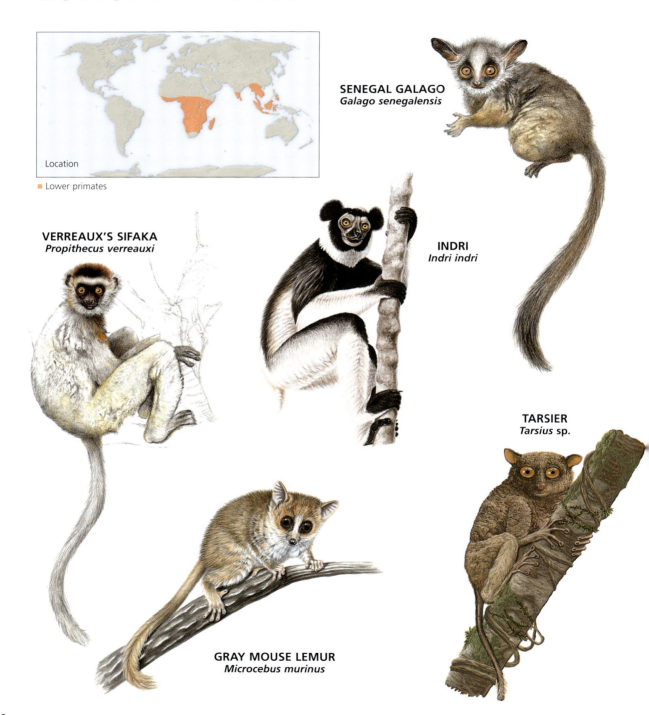

Ring-tailed lemurs live in troops of up to 20 individuals.

Black and white rings on tail

Their diet consists of foraged fruit and insects.

Troop males use horny spurs on their wrists to mark their territory.

LOWER PRIMATE ADAPTATIONS

An enlarged toilet claw, used for grooming, is only on the second digit.

Dental comb on lower jaw

RING-TAILED LEMUR
Lemur catta

DENTAL COMB
Used in mutual grooming

Lower incisors and canines

Lower molar

107

Lower Primates

Higher Primates

NEW WORLD
Woolly monkey

Larger species have a prehensile tail that acts as a fifth limb.

Broad septum with nostrils that face sideways

Thumbs are not highly opposable to other fingers.

OLD WORLD
Vervet monkey

Narrow septum with nostrils that face downward

Thumbs on forefeet and hindfeet are highly opposable (able to grasp objects easily).

Tail not prehensile

SPIDER MONKEY

Spiders monkeys move through the trees using their extended limbs or swinging from their prehensile tail.

HIGHER PRIMATE FACES

Greater white-nosed monkey

Japanese macaque

Yellow baboon

Tonkin snub-nosed monkey

LAR GIBBON

Using their bodies as a pendulum, lar gibbons swing hand-over-hand in an energy-saving process called brachiation. They can travel up to 10 feet (3 m) each swing.

Ball-and-socket joints in the wrists increase mobility.

African Old World Monkeys

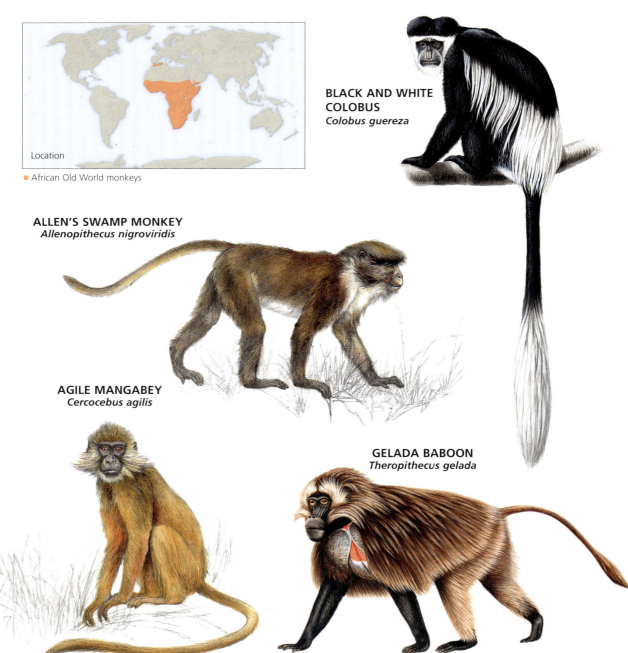

Asian Old World Monkeys

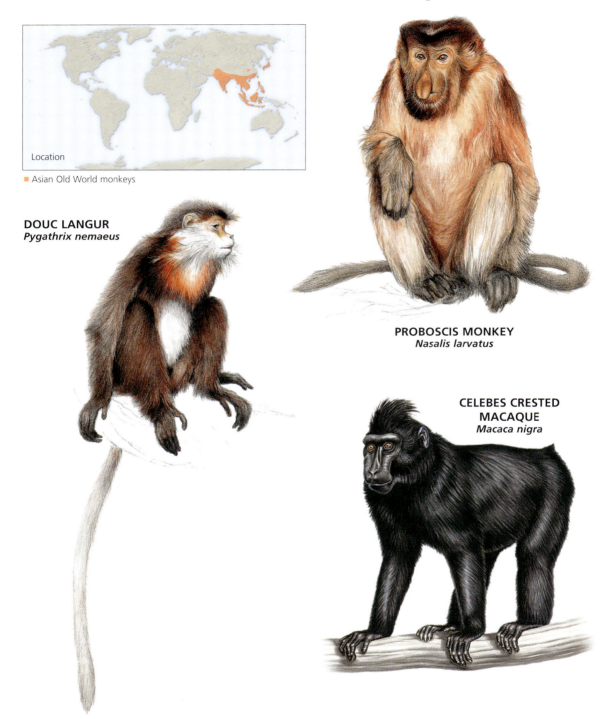

Location
■ Asian Old World monkeys

DOUC LANGUR
Pygathrix nemaeus

PROBOSCIS MONKEY
Nasalis larvatus

CELEBES CRESTED MACAQUE
Macaca nigra

New World Monkeys

Location
- New World monkeys

MIDAS TAMARIN
Saguinus midas

BROWN CAPUCHIN
Cebus apella

NORTHERN NIGHT MONKEY
Aotus trivirgatus

RED HOWLER MONKEY
Alouatta seniculus

MURIQUI
Brachyteles arachnoides

COTTON-TOP TAMARINS
Saguinus oedipus
Cotton-top tamarins live in family troops of 10–12 individuals with a single breeding pair. Older siblings and troop males help raise and carry the pair's offspring.

RED UAKARI
Cacajao calvus

MASKED TITI
Callicebus personatus

Great Apes

Location
- Apes

MOUNTAIN GORILLA
Gorilla gorilla beringei

WESTERN LOWLAND GORILLA
Gorilla gorilla gorilla

CHIMPANZEE
Pan troglodytes

Head of male orangutan

ORANGUTAN
Pongo pygmaeus

BONOBO
Pan paniscus

Gorillas

SKULL COMPARISON

Lower primate

Lemur skull

Higher primate

Gorilla skull

Human skull

FAMILY GROUPS
Mountain gorillas live in closely knit family groups of related females in the mountain forests of Africa.

A large and dominant silverback male leads and protects the group.

Gorillas build nests in the trees each night for warmth and safety.

All apes have a fully opposable thumb that is used to grasp objects.

Gorillas are plant eaters (herbivores) and spend most of the day eating.

Carnivores

EVOLUTION OF CARNIVORES

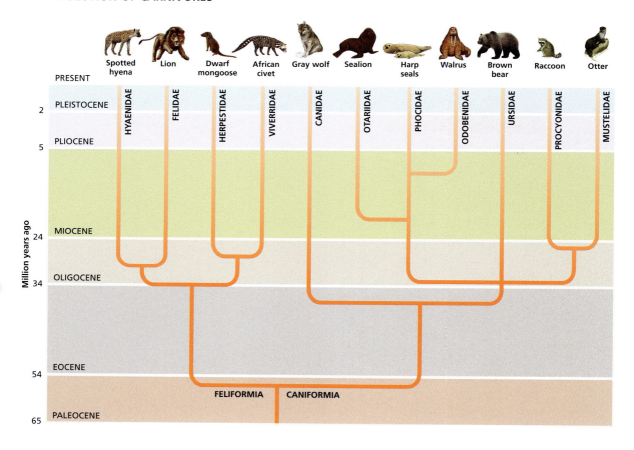

DENTAL ADAPTATIONS

■ Slicing teeth ■ Grinding teeth

Cat: highly carnivorous

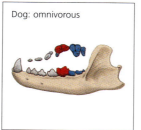

Dog: omnivorous

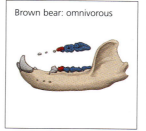

Brown bear: omnivorous

Giant panda: herbivorous

Wolves, Dogs and Foxes

GRAY WOLF PACK

Wolves are efficient hunters who work in a group to exhaust their prey.

Wolf cubs stay with their birth pack until they mature physically.

Wolves communicate with howls, body language and facial expressions.

Wolf packs enforce a strict social hierarchy within the group.

Gray wolf

Female red fox with pups

African wild dogs fighting

WOLF TAILS

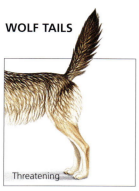
Threatening

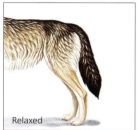
Relaxed

Fearful

Submissive

Ready to attack

DOMESTIC DOG

DOMESTIC DOG SKELETON

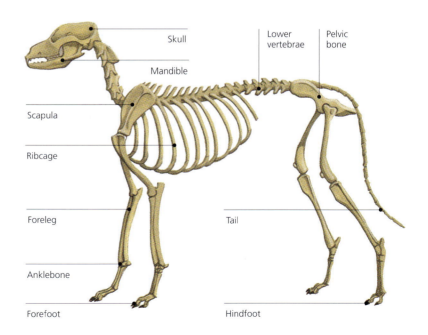

Skull
Mandible
Lower vertebrae
Pelvic bone
Scapula
Ribcage
Foreleg
Tail
Anklebone
Forefoot
Hindfoot

Wolves, Dogs and Foxes

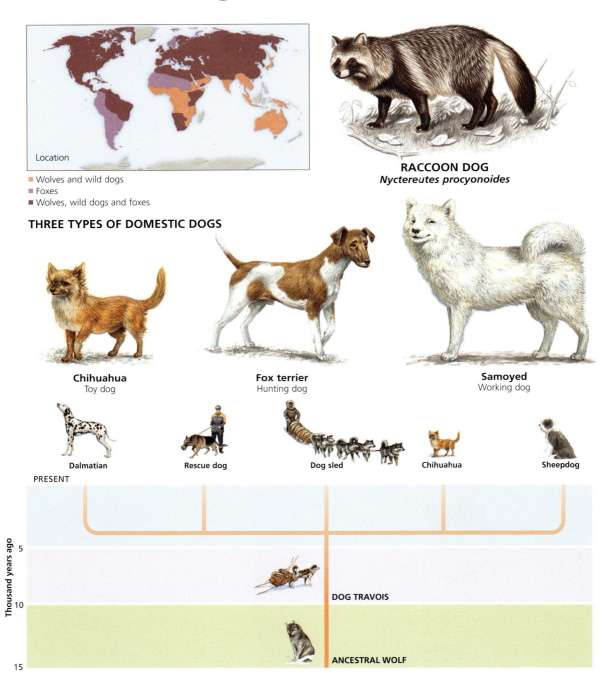

Location
- Wolves and wild dogs
- Foxes
- Wolves, wild dogs and foxes

RACCOON DOG
Nyctereutes procyonoides

THREE TYPES OF DOMESTIC DOGS

Chihuahua
Toy dog

Fox terrier
Hunting dog

Samoyed
Working dog

Dalmatian | Rescue dog | Dog sled | Chihuahua | Sheepdog

PRESENT
Thousand years ago
5
10
15

DOG TRAVOIS

ANCESTRAL WOLF

MANED WOLF
Chrysocyon brachyurus

GRAY WOLF
Canis lupus

AFRICAN WILD DOG
Lycaon pictus

BUSH DOG
Speothos venaticus

DINGO
Canis lupus dingo

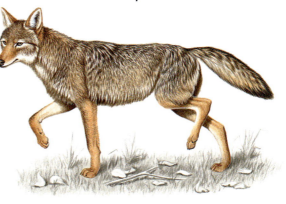

RED WOLF
Canis rufus

Wolves, Dogs and Foxes

ANATOMY OF HUNTING

- Pricked-up ears to pinpoint sound
- Forward-facing eyes to perceive depth
- Sensitive nose to detect prey
- Teeth with shearing edges to slice flesh

BAT-EARED FOX
Otocyon megalotis

GOLDEN JACKAL
Canis aureus

ARCTIC FOX
Alopex lagopus

Arctic foxes have relatively small ears and a thick pelt to prevent heat loss.

COYOTE
Canis latrans

GRAY FOX
Urocyon cinereoargenteus

KIT FOX
Vulpes macrotis

COYOTE EXPRESSIONS

Submissive

Combative

Defensive

Friendly

Playful

Bears

EVOLUTION OF BEARS

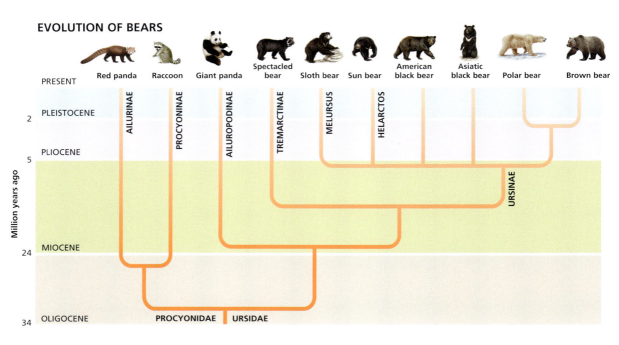

BEAR SIZES

- Sun bear
- Giant panda
- Asiatic black bear
- Sloth bear
- American black bear
- Spectacled bear
- Brown bear
- Polar bear

FOOT COMPARISON

American black bear
Long, sharp claws on forefeet

Giant panda
Short, blunt claws on forefeet

Walks flat on hindfeet

Walks on toes of hindfeet

FOREFOOT COMPARISON

Brown bear
Long claws for digging

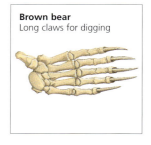

Giant panda
Modified wrist acts like digit

BROWN BEAR

TREE CLIMBERS
American black bears are agile climbers and often snooze in the safety of the treetops.

BROWN BEAR SKELETON

- Skull
- Mandible
- Scapula
- Ribcage
- Pelvic bone
- Hip
- Knee
- Anklebone
- Foreleg
- Claws

Eurasian and New World Bears

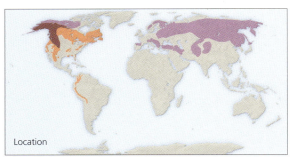
Location
- Eurasian bear (brown bear)
- New World bears
- Eurasian bear and New World bears

BROWN BEAR
Ursus arctos

AMERICAN BLACK BEAR REPRODUCTION

Winter
Hibernation and birth of young

Spring
Nursing and rearing young

Summer
Mating and intense feeding

Autumn
Delayed implantation of embryo

SPECTACLED BEAR
Tremarctos ornatus

AMERICAN BLACK BEAR
Ursus americanus

Polar Bear

Location
- Polar bear

POLAR BEAR CUBS

POLAR BEAR YEAR

1. Late February–April
Mother and pups leave den and mother replenishes fat stores.

2. March–April
Adult males travel in search of breeding females, then fight for breeding rights.

8. December–February
Except for pregnant females, polar bears actively feed throughout winter.

7. November–January
Pregnant female gives birth and nurses cubs in the den.

POLAR BEAR
Ursus maritimus

3. April–May
Mating occurs over two weeks and eggs are fertilized, but are not implanted until September.

4. May–June
At two-and-a-half years of age, young bears are weaned from their mothers.

6. August–late November
Sea ice melts and polar bears come ashore, living off their fat reserves until the sea freezes again.

5. April–July
Ringed seal pups, born in April, are easy prey for polar bears.

135

Asian Bears

Location
- Asian bears
- Giant panda

SLOTH BEAR
Melursus ursinus

ASIATIC BLACK BEAR
Ursus thibetanus

Mustelids

Location

Mustelids (badgers, weasels, otters, skunks)

AMERICAN BADGER
Taxidea taxus

EUROPEAN BADGER SETT (BURROW)
While badgers are usually solitary creatures, they build an elaborate system of communal tunnels, called a sett, with individual nesting chambers for each badger.

Entrance

Individual chamber

Individual chamber

EUROPEAN POLECAT
Mustela putorius

MARBLED POLECAT
Vormela peregusna

EURASIAN BADGER
Meles meles

HOG BADGER
Arctonyx collaris

MOLINA'S HOG-NOSED SKUNK
Conepatus chinga

HUMBOLDT'S HOG-NOSED SKUNK
Conepatus humboldtii

SKUNK DEFENSE
Mustelids have musk glands that are used for defense and marking territory.

Confronts predator face-to-face

Raises tail in warning

Rises up on forelegs

Ejects foul-smelling liquid

Mustelids

SEA OTTER
Sea otters float on their backs and use small stones to crack open sea urchins.

MARTEN
Martens are agile hunters with powerful forelimbs and a long tail. They chase prey by leaping through the trees.

EUROPEAN OTTER
Lutra lutra

GIANT OTTER
Pteronura brasiliensis

Catlike Carnivores

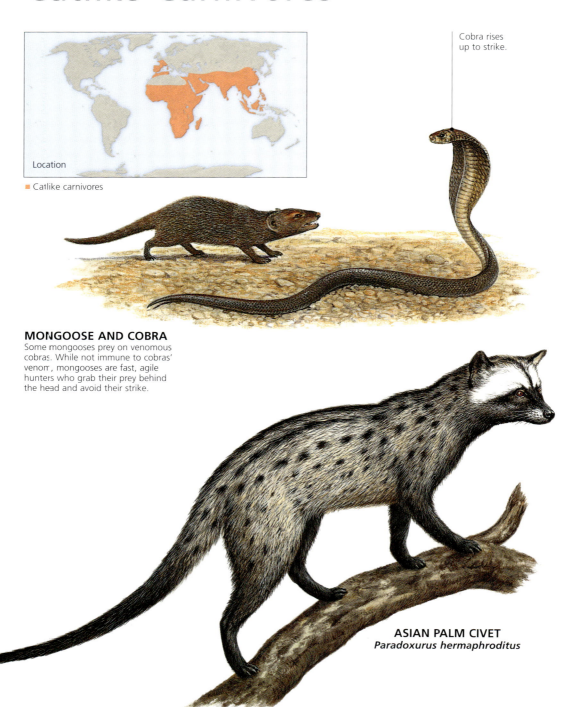

Cobra rises up to strike.

Location
■ Catlike carnivores

MONGOOSE AND COBRA
Some mongooses prey on venomous cobras. While not immune to cobras' venom, mongooses are fast, agile hunters who grab their prey behind the head and avoid their strike.

ASIAN PALM CIVET
Paradoxurus hermaphroditus

Raccoons and Hyenas

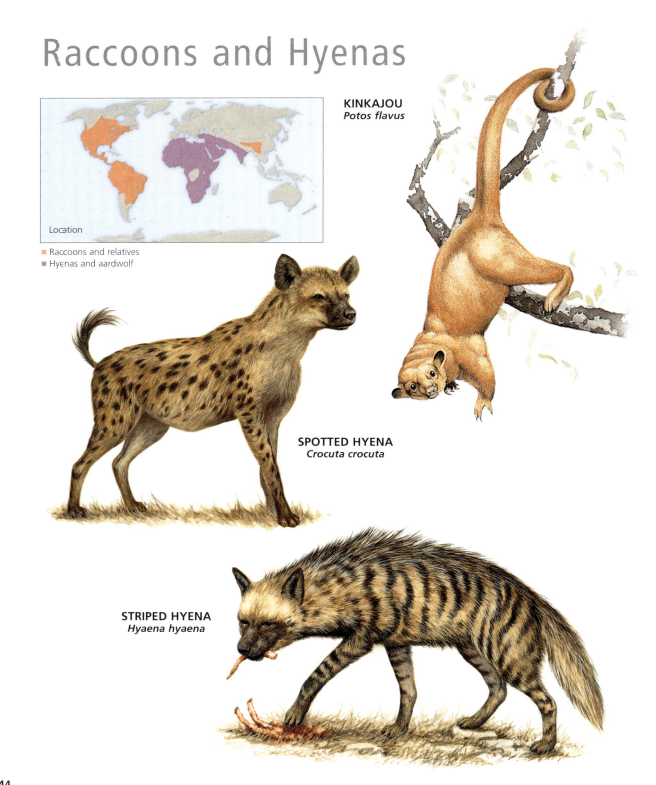

Location
- Raccoons and relatives
- Hyenas and aardwolf

KINKAJOU
Potos flavus

SPOTTED HYENA
Crocuta crocuta

STRIPED HYENA
Hyaena hyaena

Hyena Hunting Tactics

PACK HUNTING

The largest members of the hyena family, spotted hyenas are not exclusively scavengers, but are efficient pack hunters.

They are able to accelerate to speeds of over 30 miles per hour (48 km/h), and will chase their prey at this speed for more than a mile.

Other hyenas nearby join in the chase, until the gemsbok is exhausted. They work together to bring down their prey.

STEALING THE KILL

The spotted hyena's jaws are strong enough to crunch through bone; the hydrochloric acid in its stomach breaks down bone.

Rapid eaters, a pack of hyenas can demolish an adult gemsbok in as little as half an hour, this speed necessitated by threats from scavengers.

Despite their reputation as fearsome hunters, lions are just as likely to scavenge a kill from hyenas, who are then forced to abandon it.

Cats

EVOLUTION OF CATS

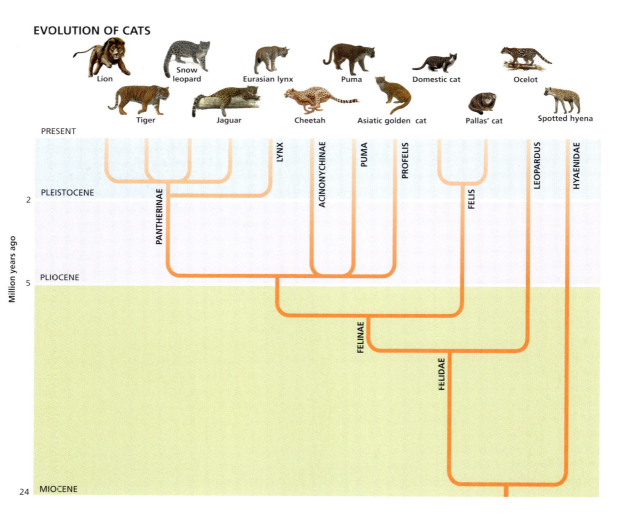

MARGAY EXPRESSIONS

Aggressive

Defensive

SPRINGLIKE CLAWS

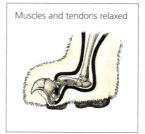

Muscles and tendons relaxed

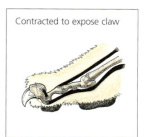

Contracted to expose claw

WILD CAT

WILD CAT SKELETON

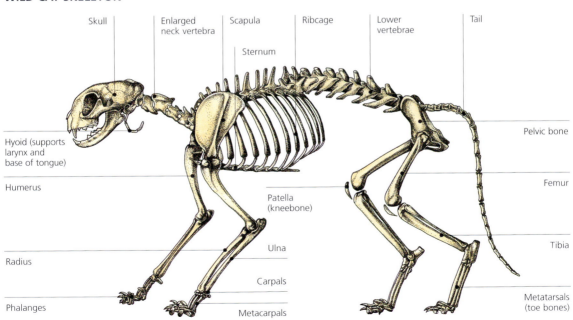

Great Cats

Location

■ Great cats

OCELOT
Leopardus pardalis

LION PRIDE

There are usually one, but often two related dominant males in the pride.

Lionesses are the primary hunters of the pride, and hunt together to provide meat for the group.

Females breed from the age of four, and mate every 20 minutes for five days when on heat.

Up to three generations of females can live in one pride, which can hold as many as 30 individuals.

Lion cubs nurse for up to six months, before weaning. Cubs are the last to feed from kills.

Great Cats

CLOUDED LEOPARD
Neofelis nebulosa

MARBLED CAT
Pardofelis marmorata

SNOW LEOPARD
Uncia uncia

LEOPARD AND GIRAFFE
Leopards often rest in trees, and will stow a kill on a branch to keep it out of reach of competitors and scavengers.

JAGUAR
Panthera onca

BLACK PANTHER
Panthera pardus
Also known as melanic leopards, panthers have a genetic mutation that raises the amount of melanin (pigment) in the skin and fur.

LEOPARD
Panthera pardus

Lynxes and Cheetah

- Lynxes, caracal and bobcat
- Lynxes, caracal, bobcat and cheetah

EURASIAN LYNX
Lynx lynx

CANADA LYNX
Lynx canadensis

BOBCAT
Lynx rufous

CARACAL
Caracal caracal

SPANISH LYNX
Lynx pardinus

CHEETAH
Acinonyx jubatus

FASTEST ON LAND
Cheetahs can run at speeds of 70 miles (110 km) per hour in short bursts of speed.

Small Cats

Location
■ Small cats

KODKOD
Oncifelis guigna

LEOPARD CAT
Prionailurus bengalensis

JUNGLE CAT
Felis chaus

FISHING CAT
Prionailurus viverrinus

ANDEAN MOUNTAIN CAT
Felis jacobita

Seals, Sealions and Walrus

FUR SEAL
Layer of blubber and thick fur to protect from cold

SEALION SKULL
Parallel rows of teeth
Flattened auditory bullae
Nasal cavity
Eye socket
Sagittal crest

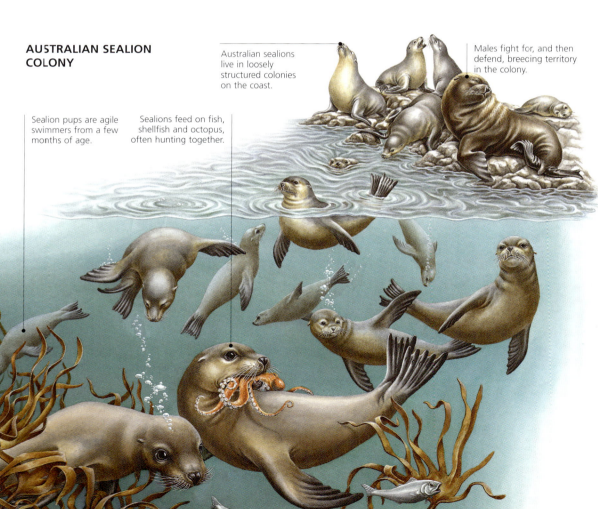

AUSTRALIAN SEALION COLONY

Australian sealions live in loosely structured colonies on the coast.

Males fight for, and then defend, breeding territory in the colony.

Sealion pups are agile swimmers from a few months of age.

Sealions feed on fish, shellfish and octopus, often hunting together.

WALRUS

WALRUS SKELETON

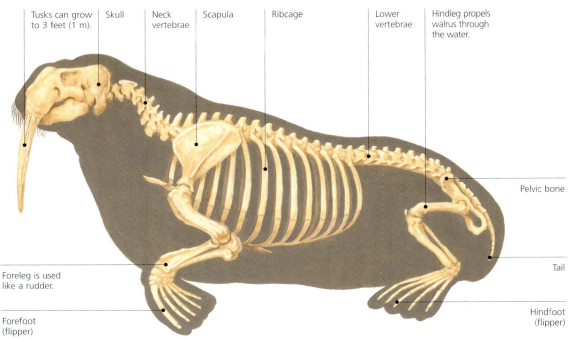

Seals, Sealions and Walrus

Location
- Seals and sealions
- Seals, sealions and walrus

BEARDED SEAL
Erignathus barbatus

HARP SEAL
Phoca groenlandica

STELLER SEALION
Eumetopias jubatus

NEW ZEALAND FUR SEAL
Arctocephalus forsteri

BAIKAL SEAL
Phoca sibirica

RIBBON SEAL
Phoca fasciata

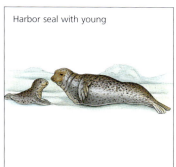

Harbor seal with young

Bull elephant seal guarding territory

Leopard seal

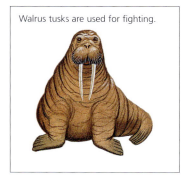

Walrus tusks are used for fighting.

Fur seal showing hind flippers

Hooded seal with inflatable skin pouch

Harp seal pups

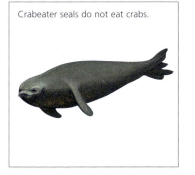

Crabeater seals do not eat crabs.

Weddell seals diving off pack ice

Whales, Dolphins and Porpoises

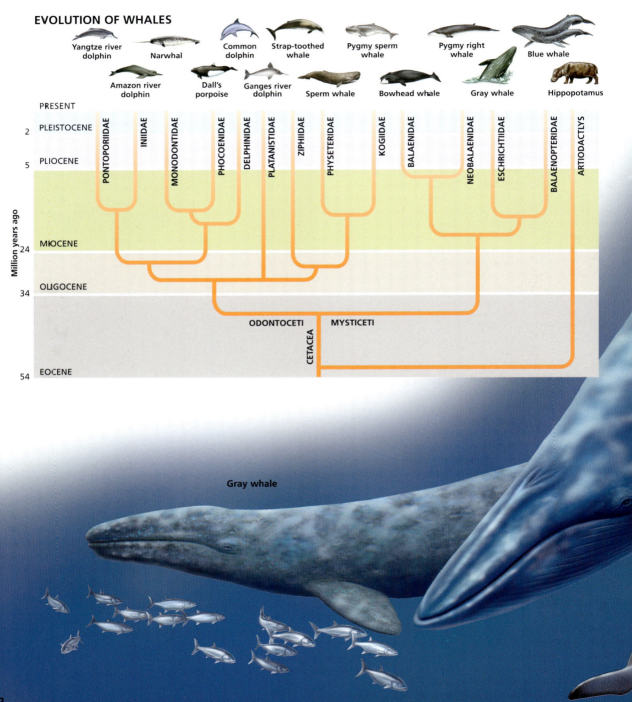

Gray whale

TYPES OF WHALES

The sei whale and the gray whale are both baleen whales, while the killer whale is a dolphin, which is a type of toothed whale.

Sei whale

Killer whale

WHALE TAILS

Common dolphin

Southern right whale

Blue whale

Humpback whale

Gray whale

Killer whale

Whale Anatomy

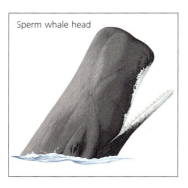
Sperm whale head

Bottlenose dolphin head

TOOTHED WHALE
Age is measured by counting rings inside the teeth.

TOOTHED WHALE
Atlantic white-sided dolphin

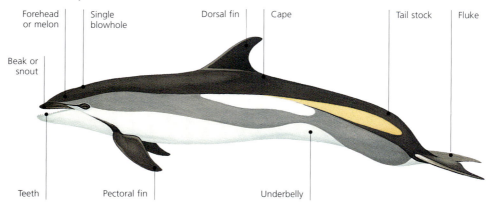

- Forehead or melon
- Single blowhole
- Dorsal fin
- Cape
- Tail stock
- Fluke
- Beak or snout
- Teeth
- Pectoral fin
- Underbelly

TOOTHED WHALE SKELETON

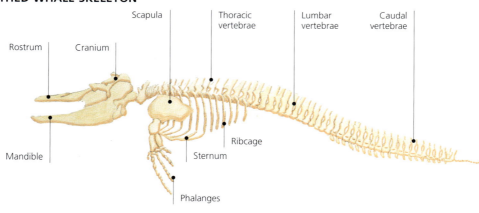

- Scapula
- Thoracic vertebrae
- Lumbar vertebrae
- Caudal vertebrae
- Rostrum
- Cranium
- Mandible
- Phalanges
- Sternum
- Ribcage

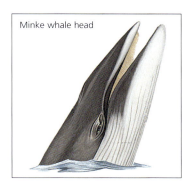
Minke whale head

Southern right whale head

BALEEN WHALE
Age is measured by counting rings inside the waxy ear plug.

BALEEN WHALE
Pygmy right whale

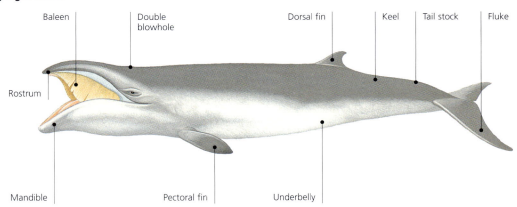

BALEEN WHALE SKELETON

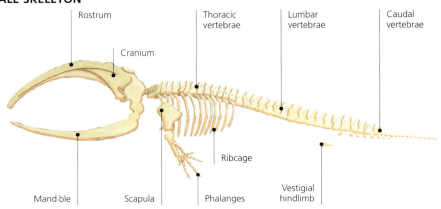

Baleen Whales

BUBBLENETTING

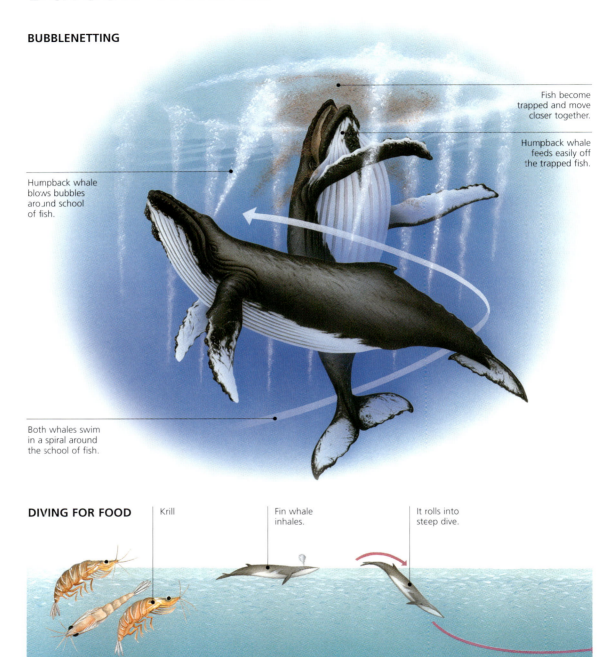

Fish become trapped and move closer together.

Humpback whale feeds easily off the trapped fish.

Humpback whale blows bubbles around school of fish.

Both whales swim in a spiral around the school of fish.

DIVING FOR FOOD — Krill — Fin whale inhales. — It rolls into steep dive.

BALEEN WHALE BEHAVIOR

Lobtailing

Pecslapping

Spyhopping

Breaching

MINKE WHALE
Although many minke whales head to warmer waters for winter breeding, others stay behind. These whales are usually not ready to breed, so save their energy for the following year.

BALEEN WHALE SPOUTS

Humpback whale

Sei whale

Right whale

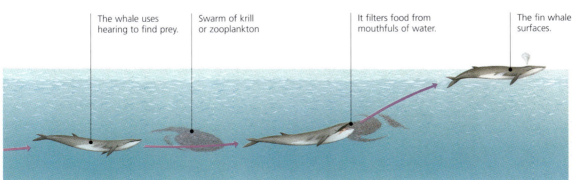

The whale uses hearing to find prey.

Swarm of krill or zooplankton

It filters food from mouthfuls of water.

The fin whale surfaces.

Baleen Whales

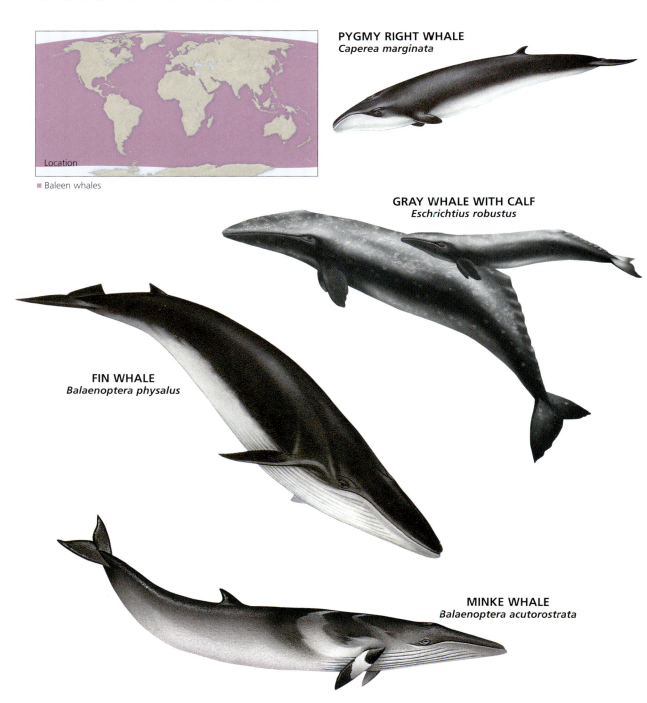

PYGMY RIGHT WHALE
Caperea marginata

GRAY WHALE WITH CALF
Eschrichtius robustus

FIN WHALE
Balaenoptera physalus

MINKE WHALE
Balaenoptera acutorostrata

Toothed Whales

DIVE DEPTHS
Some whales are able to dive to deeper depths than others in search of food. Their bodies are adapted to the intense pressure and cold of deep waters and they can hold their breath for longer periods of time.

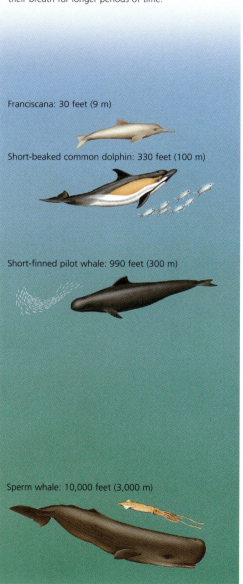

Franciscana: 30 feet (9 m)

Short-beaked common dolphin: 330 feet (100 m)

Short-finned pilot whale: 990 feet (300 m)

Sperm whale: 10,000 feet (3,000 m)

MAKING MUSIC

- Low density oil
- Blowhole
- Nasal passages and air sacs
- High density oil
- Melon
- Muscles alter shape of melon.
- "Lips" vibrate to produce sound.

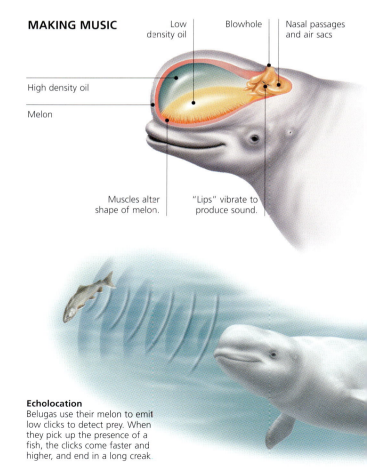

Echolocation
Belugas use their melon to emit low clicks to detect prey. When they pick up the presence of a fish, the clicks come faster and higher, and end in a long creak.

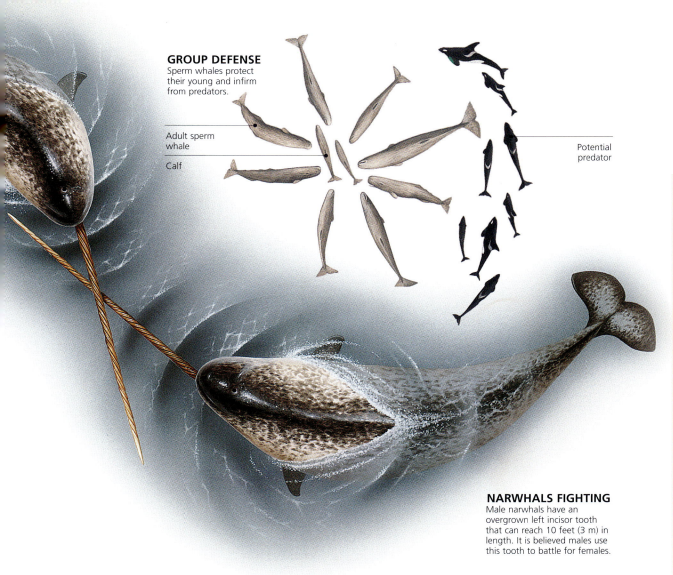

GROUP DEFENSE
Sperm whales protect their young and infirm from predators.

Adult sperm whale

Calf

Potential predator

NARWHALS FIGHTING
Male narwhals have an overgrown left incisor tooth that can reach 10 feet (3 m) in length. It is believed males use this tooth to battle for females.

Types of teeth

Narwhal

Ganges river dolphin

Sperm whale

Strap-toothed whale

Harbor porpoise

Toothed whale behavior

Touching beaks

Touching pectoral fins

171

Toothed Whales

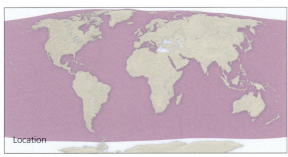
Location
■ Toothed whales

STRAP-TOOTHED WHALE
Mesoplodon layardii

SPERM WHALE DIVING | Sperm whale inhales and prepares to dive. | It dives vertically to about 1,300 feet (394 m), but some may descend to 10,000 feet (3,000 m). | It hunts in almost total darkness for prey such as jumbo flying squid and giant squid. | Giant squid

Dolphins

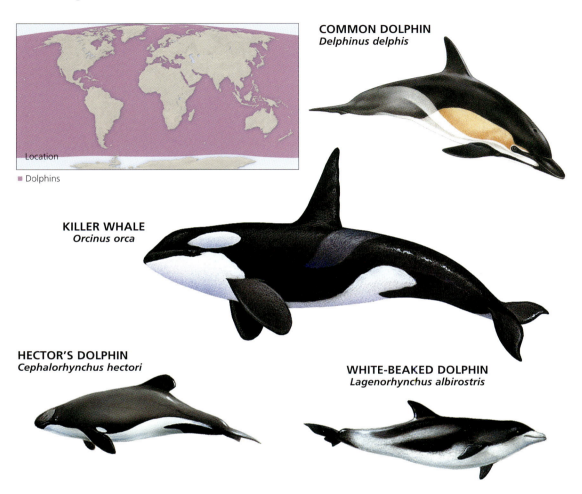

COMMON DOLPHIN
Delphinus delphis

KILLER WHALE
Orcinus orca

HECTOR'S DOLPHIN
Cephalorhynchus hectori

WHITE-BEAKED DOLPHIN
Lagenorhynchus albirostris

DOLPHIN PROPULSION

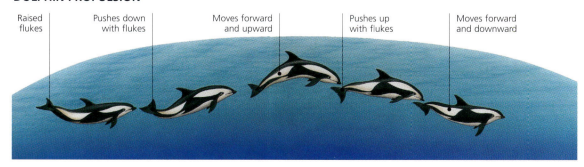

| Raised flukes | Pushes down with flukes | Moves forward and upward | Pushes up with flukes | Moves forward and downward |

River Dolphins and Porpoises

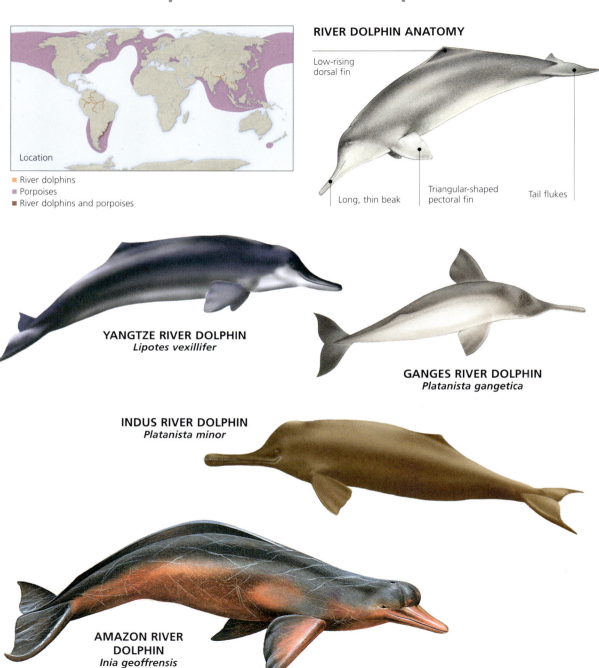

Location
- River dolphins
- Porpoises
- River dolphins and porpoises

RIVER DOLPHIN ANATOMY
- Low-rising dorsal fin
- Long, thin beak
- Triangular-shaped pectoral fin
- Tail flukes

YANGTZE RIVER DOLPHIN
Lipotes vexillifer

GANGES RIVER DOLPHIN
Platanista gangetica

INDUS RIVER DOLPHIN
Platanista minor

AMAZON RIVER DOLPHIN
Inia geoffrensis

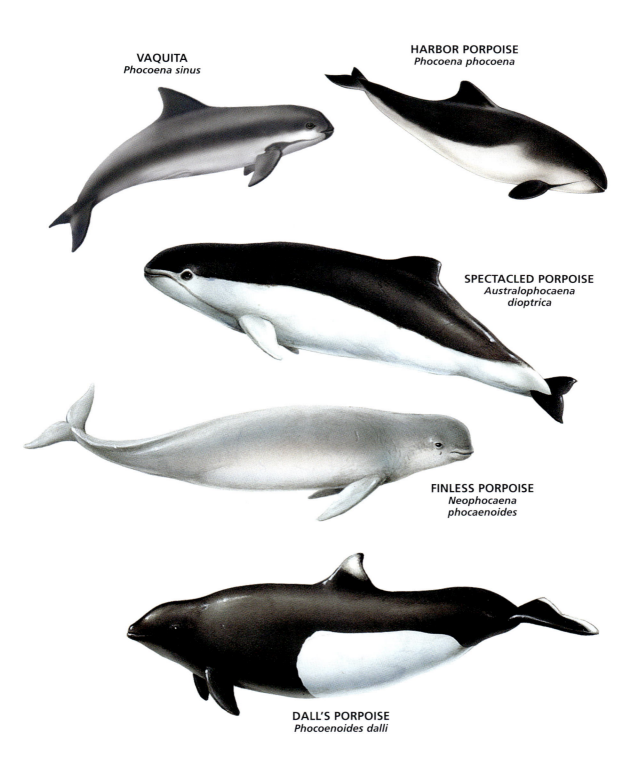

Hoofed Mammals

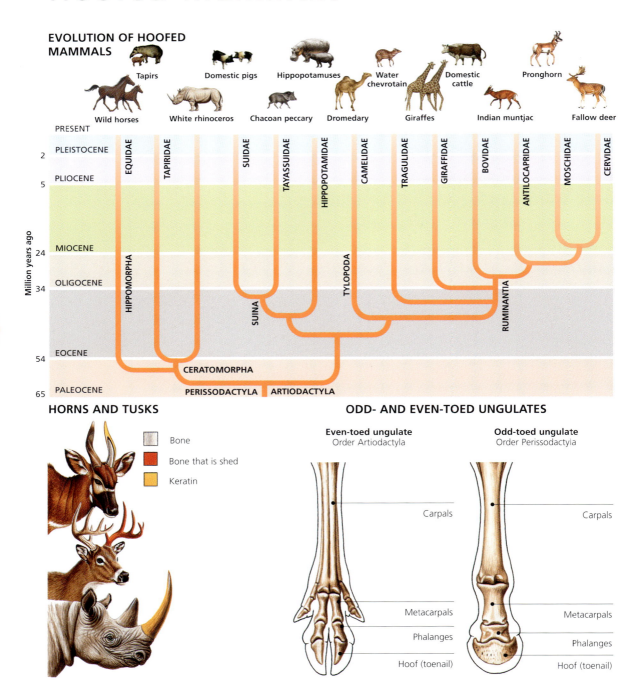

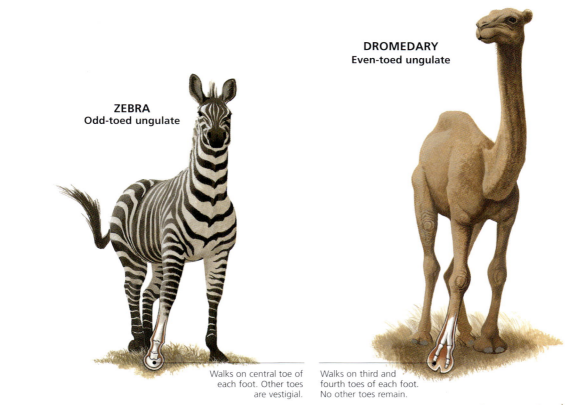

ZEBRA
Odd-toed ungulate

DROMEDARY
Even-toed ungulate

Walks on central toe of each foot. Other toes are vestigial.

Walks on third and fourth toes of each foot. No other toes remain.

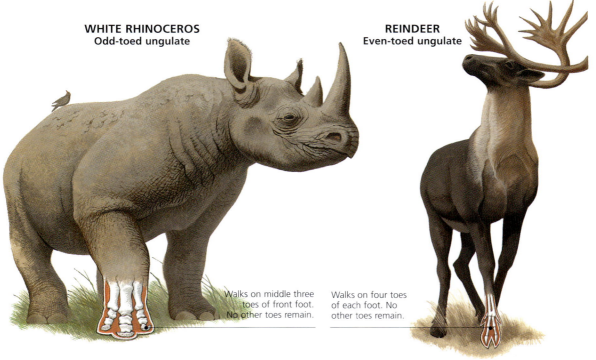

WHITE RHINOCEROS
Odd-toed ungulate

REINDEER
Even-toed ungulate

Walks on middle three toes of front foot. No other toes remain.

Walks on four toes of each foot. No other toes remain.

Hippopotamuses, Pigs and Peccaries

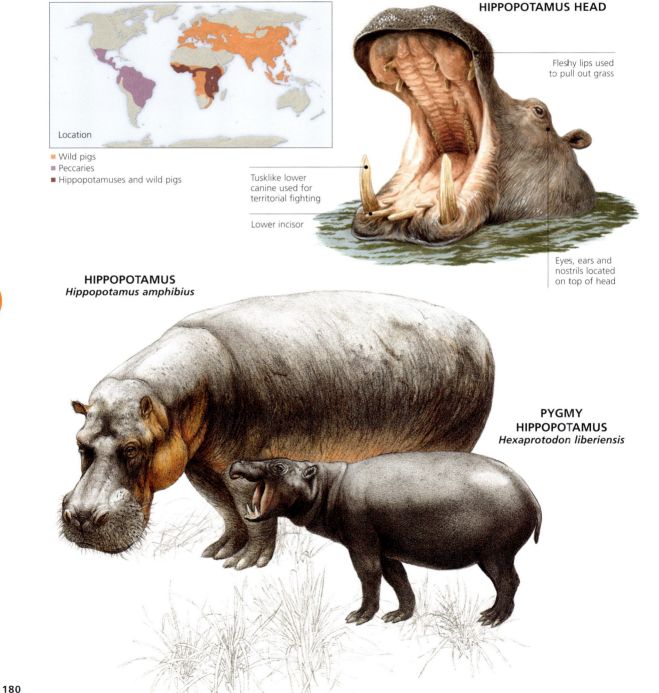

Location
- Wild pigs
- Peccaries
- Hippopotamuses and wild pigs

HIPPOPOTAMUS HEAD
- Fleshy lips used to pull out grass
- Tusklike lower canine used for territorial fighting
- Lower incisor
- Eyes, ears and nostrils located on top of head

HIPPOPOTAMUS
Hippopotamus amphibius

PYGMY HIPPOPOTAMUS
Hexaprotodon liberiensis

CHACOAN PECCARY
Catagonus wagneri

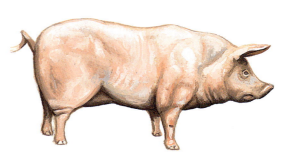
DOMESTIC PIG
Sus scrofa

BABIRUSA
Babyrousa babyrussa

RED RIVER HOG
potamochoerus porcus

BEARDED PIG
Sus barbatus

WARTHOG
Phacochoerus africanus

Old World Camels

Location

■ Bactrian camels (wild dromedary camels extinct)

CAMEL HUMP

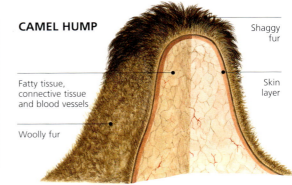

Shaggy fur

Fatty tissue, connective tissue and blood vessels

Skin layer

Woolly fur

CAMEL ANATOMY

Horny padding on hindlegs protects camels from heat.

Nostrils can be closed in sandstorms.

Camels walk two legs at a time, right following left, which causes their body to rock and sway.

Two rows of eyelashes keep out sand.

Thick fatty pads stop feet sinking.

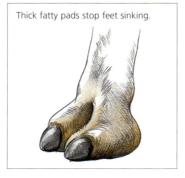

Twisted nasal passages reabsorb water.

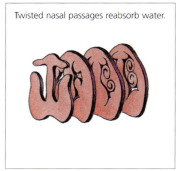

DROMEDARY
Camelus dromedarius

BACTRIAN CAMEL
Camelus bactrianus

Arabian camel-racing

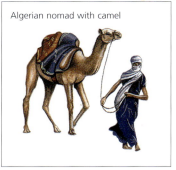

Algerian nomad with camel

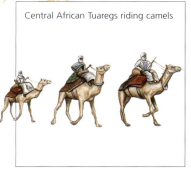

Central African Tuaregs riding camels

New World Camelids

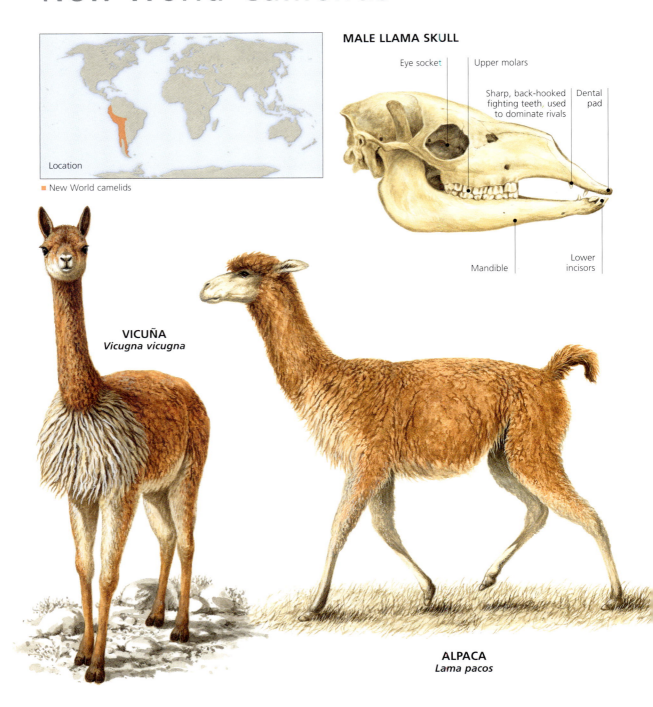

LLAMA
Lama glama

GUANACO PREPARING TO SPIT
Lama guanicoe

Domestic alpaca ready for shearing

Guancos alert for predators

Domestic llamas bred as pack animals

Giraffe, Okapi and Pronghorn

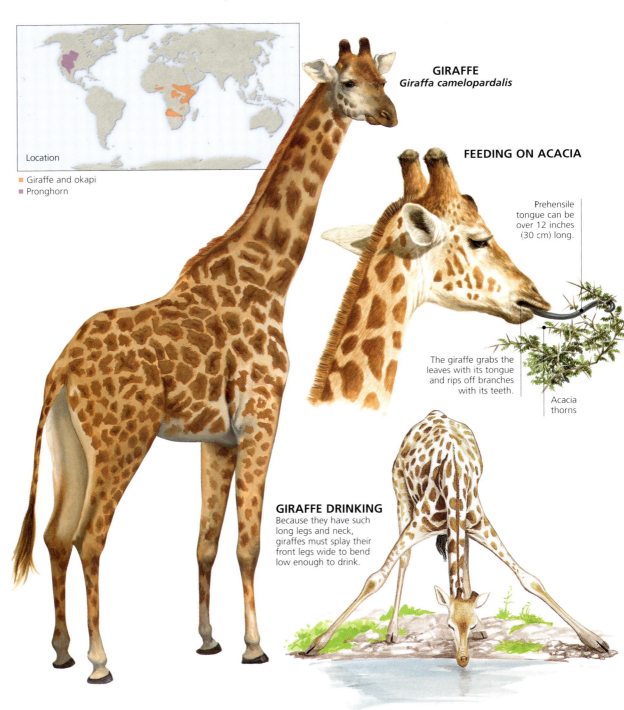

- Giraffe and okapi
- Pronghorn

GIRAFFE
Giraffa camelopardalis

FEEDING ON ACACIA

Prehensile tongue can be over 12 inches (30 cm) long.

The giraffe grabs the leaves with its tongue and rips off branches with its teeth.

Acacia thorns

GIRAFFE DRINKING
Because they have such long legs and neck, giraffes must splay their front legs wide to bend low enough to drink.

PRONGHORN TAKING FLIGHT
The fastest animal in the New World, the pronghorn flashes its white underside in warning as it bounds away from danger.

PRONGHORN
Antilocapra americana

OKAPI
Okapia johnstoni

Deer

SPREADING MOOSE ANTLERS

One year
Horns are simple spikes.

Four years
Rack expands in size and number of points.

Eight years
Mature rack demonstrates moose's suitability as a potential mate.

MALE CARIBOU FIGHTING

Male caribou fight by locking antlers during the breeding season.

Fighting males rarely become entangled by their racks of antlers.

After pushing to see which is stonger, the victor wins the right to mate with the herd's females.

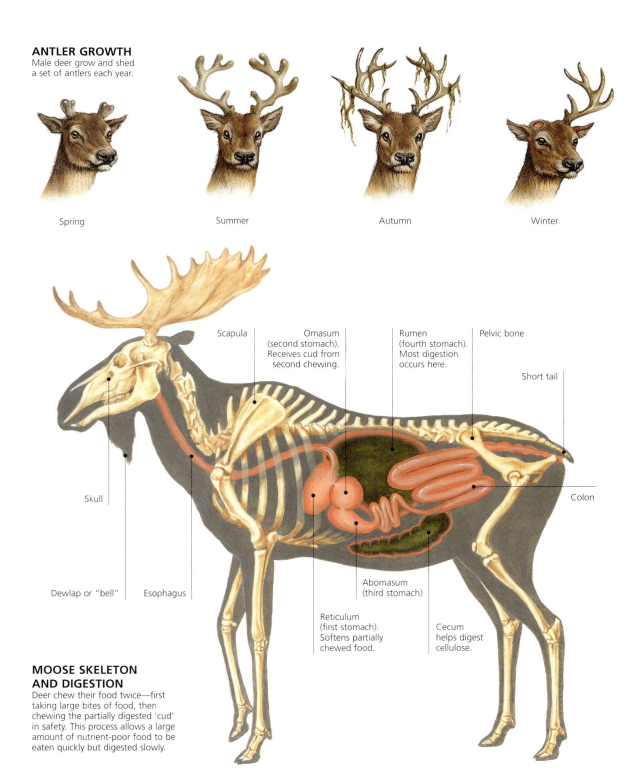

Deer

Location
- Deer, musk deer and chevrotains

PUDU
Pudu sp.

WATER CHEVROTAIN
Hyemoschus aquaticus

BARASINGHA
Cervus duvaucelii

CARIBOU
Rangifer tarandus

INDIAN MUNTJAC
Muntiacus muntjak

Bovids

GERENUKS EATING ACACIA

- Acacia shrubs
- Narrow mouth and face help the gerenuk get close to the leaves.
- Forelegs help the gerenuk to balance upright.
- Long, slender neck
- The spine curves to allow the gerenuk to stand up on hindlegs.
- Weight is distributed evenly onto the hindlegs and hooves.

MOUNTAIN SHEEP FIGHTING

Rams approach one another, horns first.   They kick and ram each other, aiming for chest and belly.   They rear up on their hindlegs.

TYPES OF BOVIDS

Barbary sheep
Goat

Gaur
True cattle

Greater kudu
Antelope

Then drop with great force into a head-on clash.　　They then present to each other.　　The rams break from fighting to feed.

Antelopes

Location
■ Antelopes, gazelles and duikers

SABLE ANTELOPE
Hippotragus niger

DUIKER
Cephalophus sp.

KLIPSPRINGER
Oreotragus oreotragus

BONGO
Tragelaphus eurycerus

BLUE WILDEBEEST
Connochaetus taurinus

Wildebeest Migration

Location
■ Wildebeest

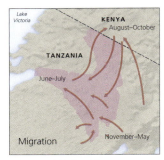

Migration
■ Serengeti National Park
■ Migration through Serengeti-Mara ecosystem

ACROSS THE SERENGETI
Each year, thousands of wildebeest make their way across the Serengeti–Mara ecosystem in Tanzania and Kenya, a journey that can be 1,900 miles (3,000 km) long. In the dry season, the wildebeest move from acacia woodlands onto the Serengeti Plain in search of water and grass. They return in the wet season.

Calves are born at the beginning of the rainy season, and stay with their mothers until eight months of age, when they leave to form peer-group herds.

Wildebeest never stay in one spot long enough to damage the environment, and as the columns of animals move forward, the soil is fertilized by their droppings.

Zebras migrate across the Serengeti at the same time, eating grasses cropped short by the wildebeest, and using the tens of thousands of animals as protection from predators.

River crossings, such as the Mara River, can be deadly, with crocodiles and other predators lying in wait for an animal to stumble or drift away from the mass.

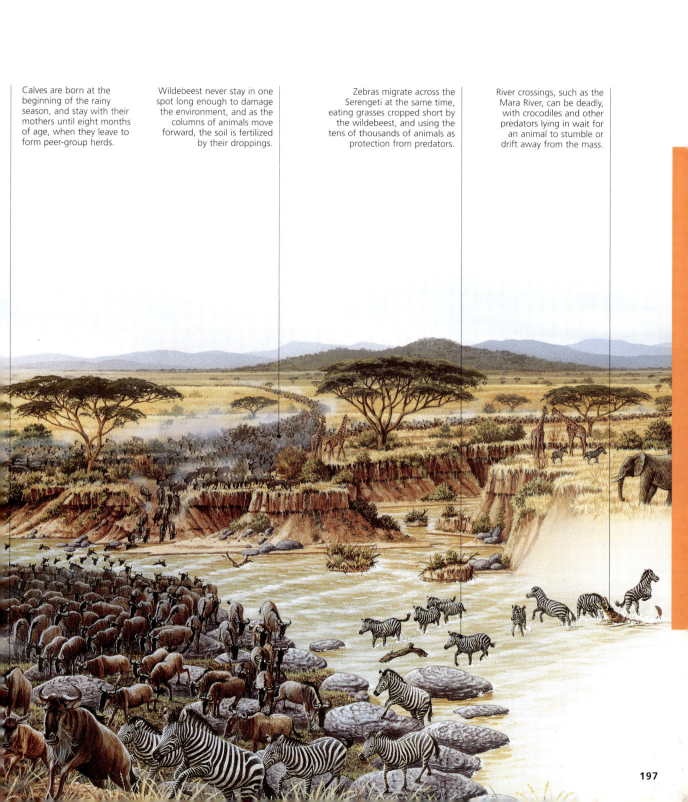

True Cattle, Goats and Sheep

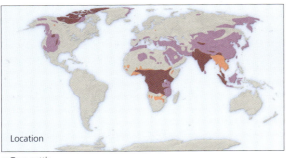

Location
- True cattle
- Goats and sheep
- True cattle, goats and sheep

YAK
Bos grunniens

MOUNTAIN ANOA
Bubalus quarlesi

AMERICAN BISON
Bison bison

EUROPEAN BISON
Bison bonasus

AFRICAN BUFFALO
Syncerus caffer

DOMESTIC BREEDS OF CATTLE

Hereford cattle

Friesian cattle

Galloway cattle

Brahmin cattle

Ankole cattle

Texas long-horned cattle

MUSKOX
Ovibos moschatus

MOUFLON
Ovis orientalis musimon

DALL'S SHEEP
Ovis dalli

BIGHORN SHEEP
Ovis canadensis

HIMALAYAN TAHR
Hemitragus jemlahicus

WILD GOAT
Capra aegagrus

MOUNTAIN GOAT
Oreamnos americanus

DOMESTIC BREEDS OF SHEEP AND GOATS

Leicester sheep

Cashmere goat

Lincoln sheep

Black-faced sheep

Merino sheep

Domestic sheep

Romney sheep

Rhinoceroses

Location
■ Rhinoceroses

RHINO DEFENSE
Groups of rhinoceroses form a circle facing outward to protect their calves.

BLACK RHINO CHARGING

Black rhinos attack animals that threaten them.

Horns made of hollow filaments similar to hair

Spotted hyena

Black rhinos use their horn to gore and toss would-be attackers.

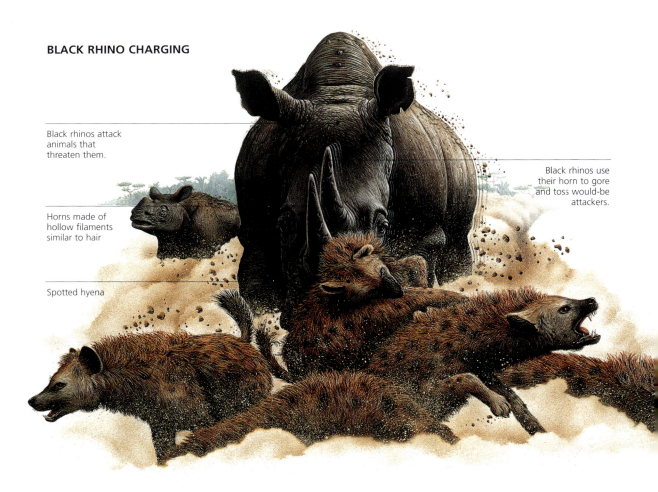

JAVAN RHINOCEROS	INDIAN RHINOCEROS	SUMATRAN RHINOCEROS	WHITE RHINOCEROS	BLACK RHINOCEROS
Rhinoceros sondaicus	*Rhinoceros unicornis*	*Dicerorhinus sumatrensis*	*Ceratotherium simum*	*Diceros bicornis*

RHINO COMPARISON

Indian rhinoceros
Asian one-horned

Black rhinoceros
African two-horned

Horses, Asses and Zebras

HORSE GALLOPING
When horses gallop, each hoof strikes the ground separately.

HORSE SKELETON AND DIGESTION
Horses chew their food once only, and have a single stomach with a long intestinal system and large, long teeth to break down tough, fibrous grasses.

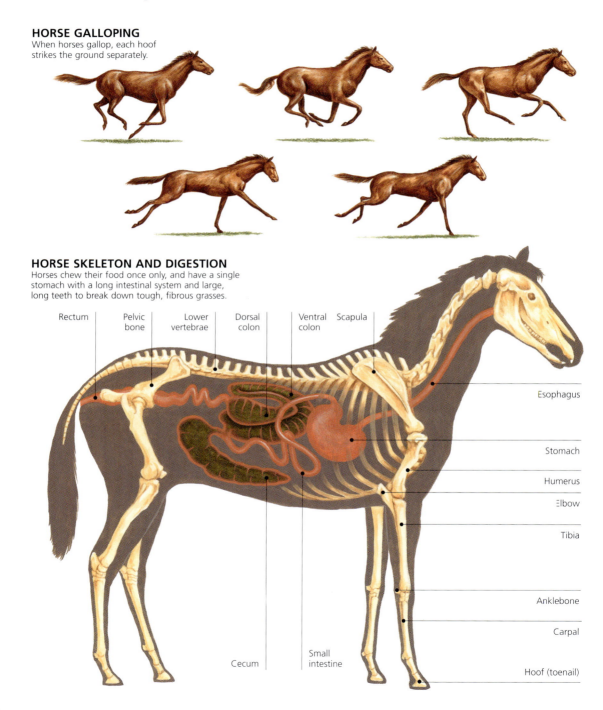

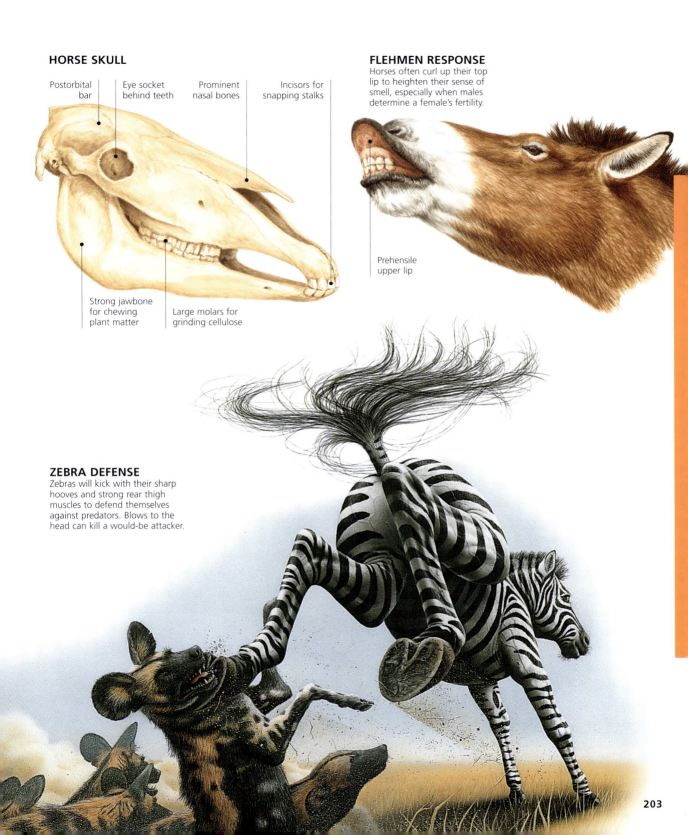

HORSE SKULL
- Postorbital bar
- Eye socket behind teeth
- Prominent nasal bones
- Incisors for snapping stalks
- Strong jawbone for chewing plant matter
- Large molars for grinding cellulose

FLEHMEN RESPONSE
Horses often curl up their top lip to heighten their sense of smell, especially when males determine a female's fertility.
- Prehensile upper lip

ZEBRA DEFENSE
Zebras will kick with their sharp hooves and strong rear thigh muscles to defend themselves against predators. Blows to the head can kill a would-be attacker.

Horses, Asses and Zebras

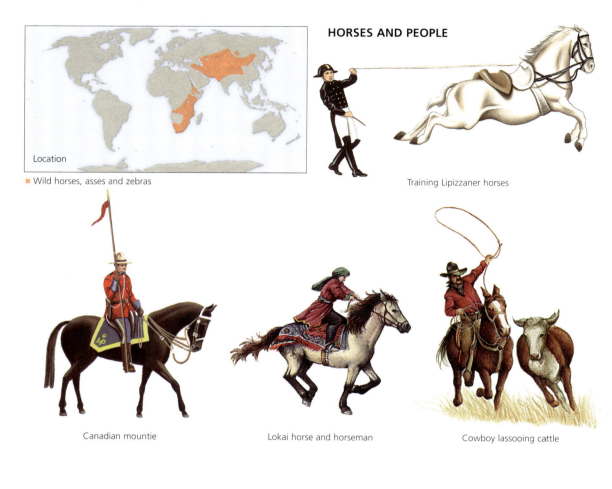

Location
■ Wild horses, asses and zebras

HORSES AND PEOPLE

Training Lipizzaner horses

Canadian mountie

Lokai horse and horseman

Cowboy lassooing cattle

Playing polo

Racing horses

Austrian sled racing

KIANG
Equus kiang

ONAGER
Equus onager

ASS
Equus asinus

HORSE
Equus caballus

TARPAN
Equus caballus gmelini

ZEBRA
Equus sp.

DOMESTIC HORSES

Feral horses

Plains pony

Arabian horse

Aardvark, Hyraxes and Tapirs

Location
- Aardvark and hyraxes
- Tapirs

YELLOW-SPOTTED HYRAX
Heterohyrax brucei

ROCK HYRAXES
Procavia capensis

AARDVARK
Orycteropus afer

AARDVARK ANATOMY

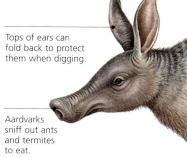

Tops of ears can fold back to protect them when digging.

Aardvarks sniff out ants and termites to eat.

Elephants

ELEPHANT SKIN

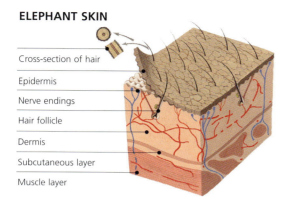

- Cross-section of hair
- Epidermis
- Nerve endings
- Hair follicle
- Dermis
- Subcutaneous layer
- Muscle layer

ELEPHANT FOOT
Elephants walk on toes of feet.   Fibrous pads support the toes.

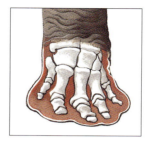

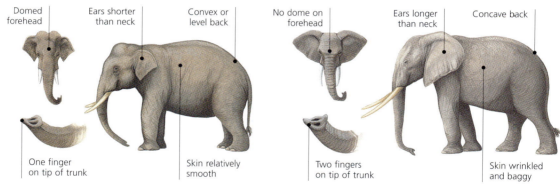

ASIAN ELEPHANT
- Domed forehead
- Ears shorter than neck
- Convex or level back
- One finger on tip of trunk
- Skin relatively smooth

AFRICAN ELEPHANT
- No dome on forehead
- Ears longer than neck
- Concave back
- Two fingers on tip of trunk
- Skin wrinkled and baggy

ELEPHANT GROWTH

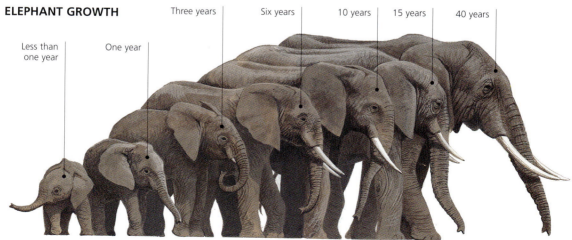

Less than one year • One year • Three years • Six years • 10 years • 15 years • 40 years

AFRICAN ELEPHANT

AFRICAN ELEPHANT SKELETON

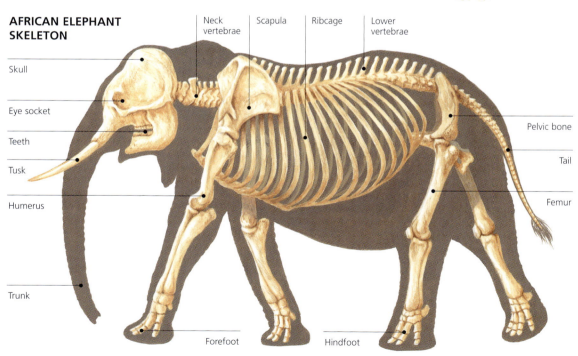

Elephants

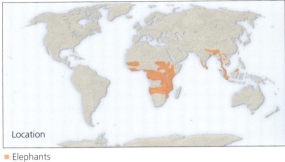

ELEPHANT TUSKS

Tusks have cavities containing nerve endings, which are pressure-sensitive

AFRICAN BUSH ELEPHANTS
Loxodonta africana

SUMATRAN ASIAN ELEPHANTS
Elephas maximus sumatranus

SRI LANKAN ASIAN ELEPHANTS
Elephas maximus maximus

MAINLAND ASIAN ELEPHANTS
Elephas maximus indicus

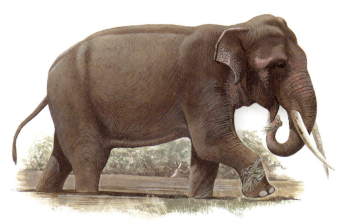

FOREST AFRICAN ELEPHANTS
Loxodonta cyclotis

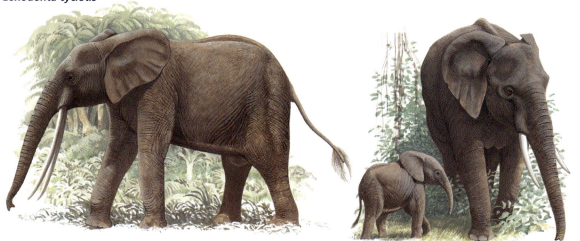

Dugong and Manatees

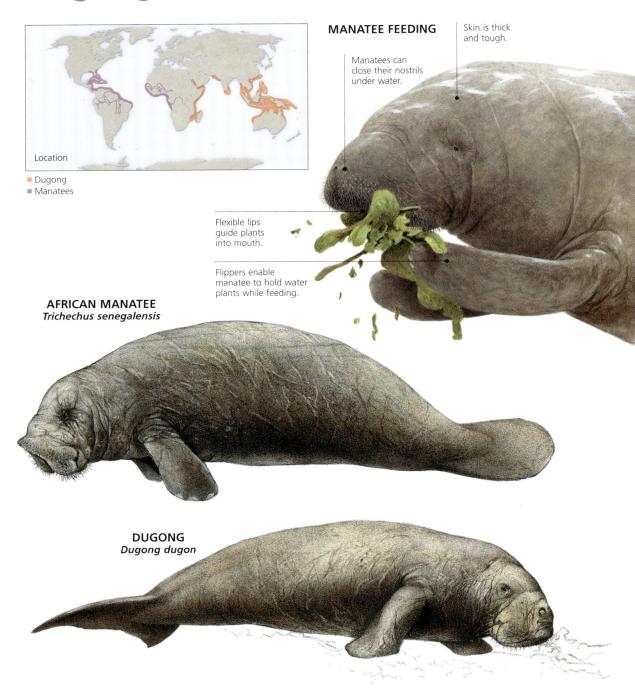

Location
- Dugong
- Manatees

MANATEE FEEDING

Skin is thick and tough.

Manatees can close their nostrils under water.

Flexible lips guide plants into mouth.

Flippers enable manatee to hold water plants while feeding.

AFRICAN MANATEE
Trichechus senegalensis

DUGONG
Dugong dugon

Elephant Shrews

Location
■ Elephant shrews

ELEPHANT SHREW HEAD

Ear

Trunklike nose

Large, white-rimmed eyes

ROCK ELEPHANT SHREW HABITAT
Elephantulus myurus

Rock elephant shrews nest in the crevices of rock overhangs to protect themselves from predators and the elements.

They have large eyes and a long, sensitive and mobile snout, which is used to find insects.

Rock elephant shrews are generally soft gray-brown in color, which acts as camouflage from predators.

SHORT-EARED ELEPHANT SHREW
Macroscelides proboscideus

GOLDEN-RUMPED ELEPHANT SHREW
Rhynchocyon chrysopygus

RUFOUS ELEPHANT SHREW
Elephantulus rufescens

NORTH AFRICAN ELEPHANT SHREW
Elephantulus rozeti

FOUR-TOED ELEPHANT SHREW
Petrodromus tetradactylus

Rodents

EVOLUTION OF RODENTS

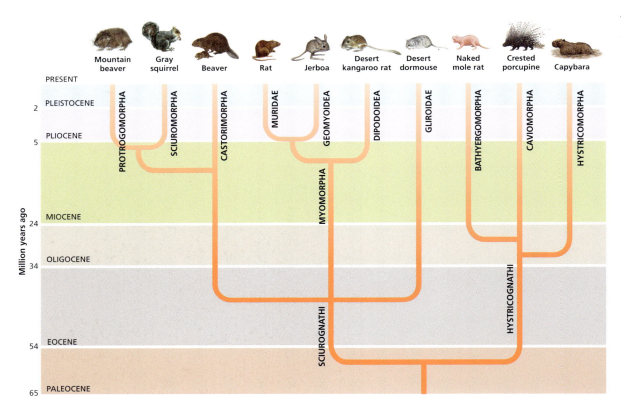

RODENT SKULL

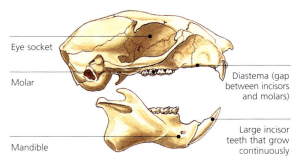

RODENT HEAD

RODENT JAW MUSCLES

■ Lateral jaw muscles ■ Deep jaw muscles

Squirrel-like rodents

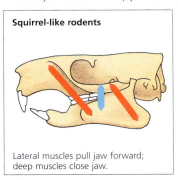

Lateral muscles pull jaw forward; deep muscles close jaw.

Mouselike rodents

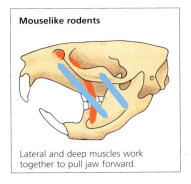

Lateral and deep muscles work together to pull jaw forward.

Cavylike rodents

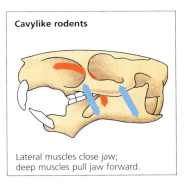

Lateral muscles close jaw; deep muscles pull jaw forward.

EURASIAN RED SQUIRREL
Squirrel-like rodent

PREHENSILE-TAILED PORCUPINE
Cavylike rodent

MEXICAN SPINY POCKET MOUSE
Mouselike rodent

LEAST CHIPMUNK
Squirrel-like rodent

MOUNTAIN BEAVER
Squirrel-like rodent

Squirrel-like Rodents

Location
■ Squirrel-like rodents

PRAIRIE DOGS
Cynomys sp.

PRAIRIE DOG COMMUNITY

Prairie dogs greet each other with kissing ceremonies.

Guards whistle an alarm when in danger.

Ferrets can crawl through the tunnels in search of prey.

Females build grass nests safe underground.

Members of different coteries greet each other by sniffing.

Prairie dogs live in social units called coteries. A number of interconnected coteries form a town.

Snakes and burrowing owls often take over unused burrows.

EASTERN FOX SQUIRREL
Sciurus niger

DOUGLAS' SQUIRREL
Tamiasciurus douglasii

MARMOT
Marmota sp.

WOODCHUCK
Marmota monax

GRAY SQUIRREL
Sciurus sp.

ABERT'S SQUIRREL
Sciurus aberti

EASTERN CHIPMUNK
Tamias striatus

Squirrel-like Rodents

BLACK GIANT SQUIRREL
Ratufa bicolor

DELMARVA FOX SQUIRREL
Sciurus niger cenerus

AMERICAN RED SQUIRREL
Tamiasciurus hudsonicus

LEAST CHIPMUNK
Tamias minimus

HOARY MARMOT
Marmota caligata

COLUMBIAN GROUND SQUIRREL
Spermophilus columbianus

Beavers

Location
■ Beavers and mountain beavers

AMERICAN BEAVER AND KIT
Castor canadensis

BEAVER DAM AND LODGE

Beavers eat bark and use their strong incisor teeth to chew through tree trunks.

After felling trees and saplings for cross-braces, beavers carry smaller logs in their mouths.

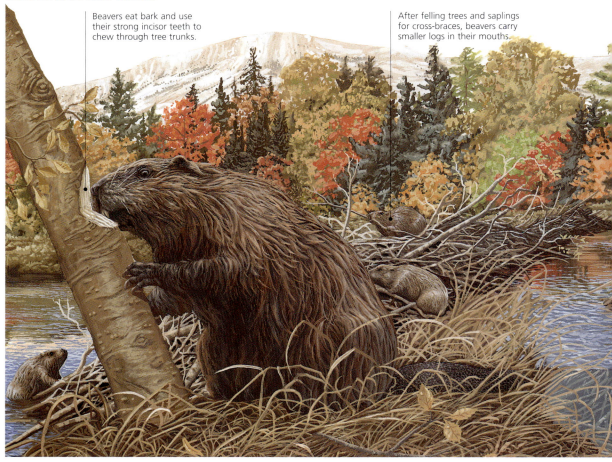

EURASIAN BEAVER
Castor fiber

DAM CONSTRUCTION
Heavier logs run crosswise.

Smaller logs run lengthwise.

LODGE ENTRANCE
Before damming

After damming

Beaver lodge

The lodge's entrance is underwater to protect it from predators.

Inside the lodge, temperatures are much more constant than in the external climate.

Mouselike Rodents

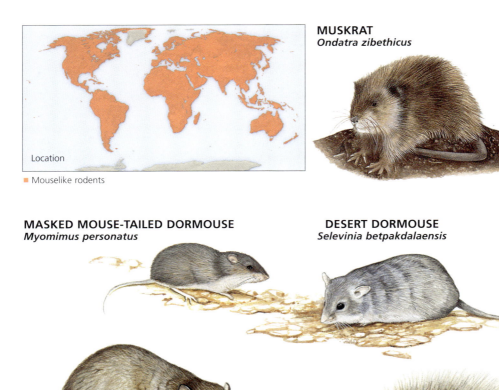

MUSKRAT
Ondatra zibethicus

Location
■ Mouselike rodents

MASKED MOUSE-TAILED DORMOUSE
Myomimus personatus

DESERT DORMOUSE
Selevinia betpakdalaensis

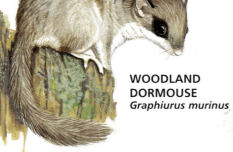

WOODLAND DORMOUSE
Graphiurus murinus

CRESTED RAT
Lophiomys imhausi

LEMMING
Lemmus sp.

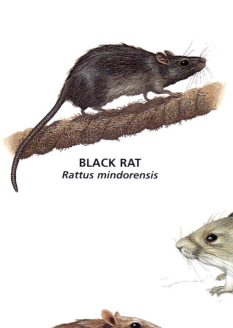
BLACK RAT
Rattus mindorensis

HOUSE RAT
Rattus rattus

LONG-TAILED POCKET MOUSE
Chaetodipus formosus

GERBIL
Gerbillus sp.

JERBOA
Jaculus sp.

BANK VOLE
Clethrionomys glareolus

GRASSHOPPER MOUSE
Onychomys sp.

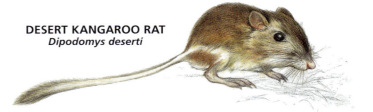
DESERT KANGAROO RAT
Dipodomys deserti

Cavylike Rodents

Location
■ Cavylike rodents

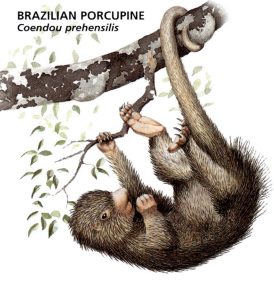
BRAZILIAN PORCUPINE
Coendou prehensilis

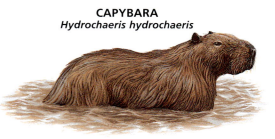
CAPYBARA
Hydrochaeris hydrochaeris

BRAZILIAN AGOUTI
Dasyprocta leporina

CRESTED PORCUPINE
Hystrix cristata

MALAYAN PORCUPINE
Hystrix brachyura

CORURO
Spalacopus cyanus

GUINEA PIG
Cavia aperea

CHILEAN ROCK RAT
Aconaemys fuscus

PLAINS VISCACHA
Lagostomus maximus

CHINCHILLA
Chinchilla lanigera

PATAGONIAN MARA
Dolichotis patagonum

SPINY RAT
Family *Echimyidae*

Rabbits, Hares and Pikas

BLACK-LIPPED PIKA
Ochotona curzoniae
Pikas are short-legged and short-eared relatives of rabbits and hares. Mainly herbivorous, they live on the steppes and plains of North America and continental Asia.

RABBIT ANATOMY

- Long ears to detect approach of predators and to keep cool
- Curved skull
- Eyes at side of head to locate predators from above and behind
- Continuously growing incisors

RABBIT WARREN

- Warrens have more than one entrance.
- Females can give birth to young as many as six times per year.
- Rabbit kittens remain in grass-lined nests until weaned at four weeks.
- Rabbit warrens are dug just large enough for an adult to squeeze through.

SNOWSHOE HARE ADAPTATIONS

Snowshoe hares eat treebark in winter to survive.

White winter coat

Comparatively short ears minimize heat loss.

Huge feet covered in long, stiff hairs distribute weight.

Brown summer coat

Summer diet of green vegetation and berries

RABBIT ADAPTATIONS

Arctic hare
Small extremities and thick pelt reduce the body's surface area to conserve heat.

Antelope jackrabbit
Long extremities and small body core enlarge its surface area to dispel heat.

Black-tailed jackrabbit
Hundreds of blood vessels in the ears radiate heat away from the body.

Rabbits, Hares and Pikas

Location
- Rabbits and hares
- Rabbits, hares and pikas

SUMATRAN RABBIT
Nesolagus netscheri

RYUKYU RABBIT
Pentalagus furnessi

VOLCANO RABBIT
Romerolagus diazi

BRUSH RABBIT
Sylvilagus bachmani

RIVENNE RABBIT
Bunolagus monticularis

PYGMY RABBIT
Brachylagus idahoensis

BIRDS

INTRODUCING BIRDS

Classifying Birds	236
Bird Characteristics	238
Evolution of Birds	240
Bird Anatomy	242

FLIGHTLESS BIRDS

Flightless Birds	244

PENGUINS

Penguins	246

GREBES AND DIVERS

Grebes and Divers	248

ALBATROSSES AND PETRELS

Albatrosses and Petrels	250

PELICANS

Pelicans	252

HERONS

Herons	254

FLAMINGOS

Flamingos	258

WATERFOWL

Waterfowl	260

BIRDS OF PREY

Birds of Prey	264
Eagles, Kites and Hawks	266
Falcons	270
Secretary Bird and Vultures	272

GAMEBIRDS

Gamebirds	274

CRANES

Cranes	278

WADERS AND SHOREBIRDS

Waders and Shorebirds	282
Waders	284
Shorebirds and Seabirds	286

PIGEONS AND SANDGROUSE

Pigeons and Sandgrouse	288

PARROTS

Parrots	290

CUCKOOS AND TURACOS

Cuckoos and Turacos	294

OWLS

Owls	296

NIGHTJARS AND FROGMOUTHS

Nightjars and Frogmouths	298

HUMMINGBIRDS AND SWIFTS

Hummingbirds and Swifts	300

MOUSEBIRDS AND TROGONS

Mousebirds and Trogons	304

KINGFISHERS

Kingfishers	306

WOODPECKERS AND TOUCANS

Woodpeckers and Toucans	310

PASSERINES

Passerines	314
Broadbills, Pittas, Ovenbirds and Tyrant Flycatchers	316
Lyrebirds, Scrub-birds, Larks, Wagtails and Swallows	318
Cuckooshrikes, Bulbuls, Leafbirds, Shrikes, Vangas and Waxwings	320
Mockingbirds, Accentors, Dippers, Thrushes, Babblers and Wrens	322
Warblers, Flycatchers, Fairywrens, Logrunners and Monarchs	324
Tits, Nuthatches, Treecreepers and Honeyeaters	326
Buntings, Tanagers, Vireos and Wood Warblers	328
Finches	330
Starlings, Magpie-larks and Icterids	332
Bowerbirds and Birds-of-Paradise	334
Crows and Jays	336

Classifying Birds

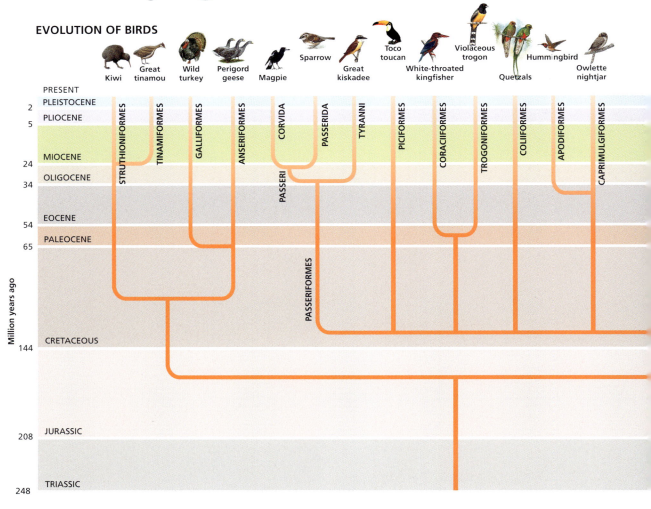

EVOLUTION OF BIRDS

Male (top) and female gambel's quails

Bonellis eagle

Painted stork

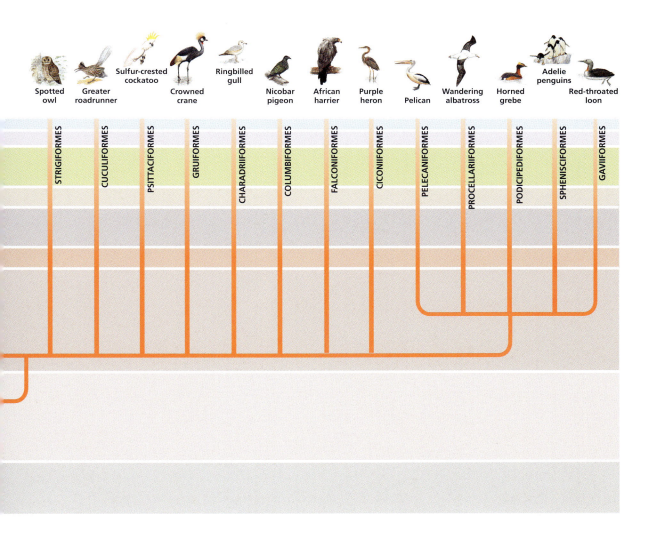

Bird Characteristics

TYPES OF FEATHERS

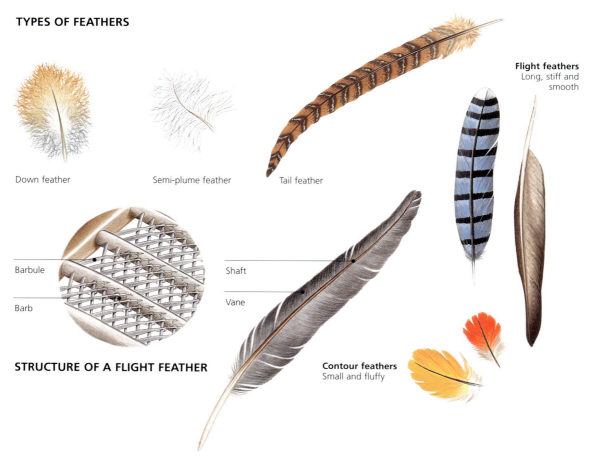

STRUCTURE OF A FLIGHT FEATHER

REPLACING FEATHERS

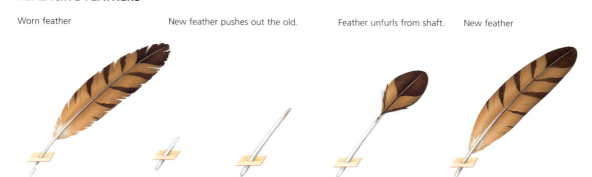

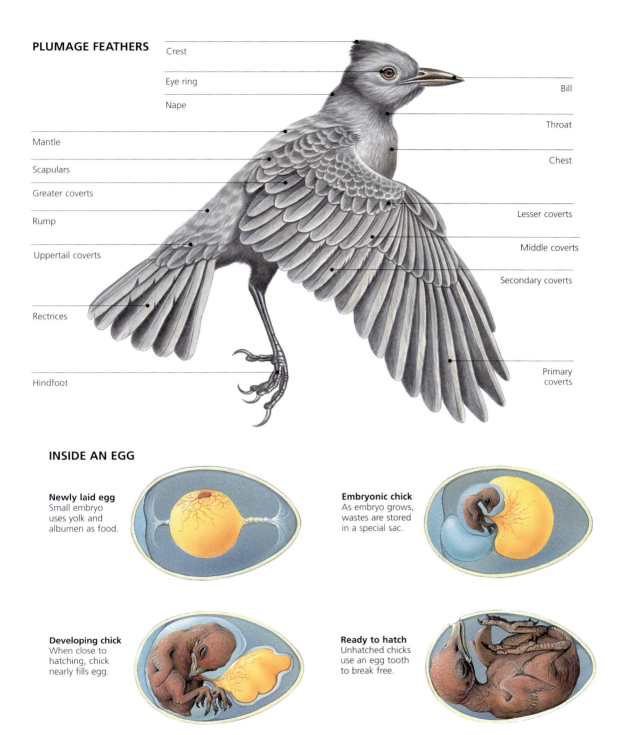

Evolution of Birds

HESPERORNIS
Ancestor to modern grebes that lived 70 million years ago.

ARCHAEOPTERYX
The earliest known bird, it lived 145 million years ago.

Waterfowl ancestor
Ichthyornis

Earliest bird
Archaeopteryx

Dinosaur ancestor
Reptilian theropod

TERROR-BIRD
At 9 feet (2.7 m), the largest of all known birds lived 1 million years ago.

EVOLUTION OF MODERN BIRDS

Modern waterfowl
Snow goose

DINORNIS MAXIMUS
Recently extinct flightless bird that lived 200–300 years ago.

TERATORNIS MERRIAMI
An ancestor of modern vultures, it lived 2 million years ago.

Bird Anatomy

AVIAN ANATOMY

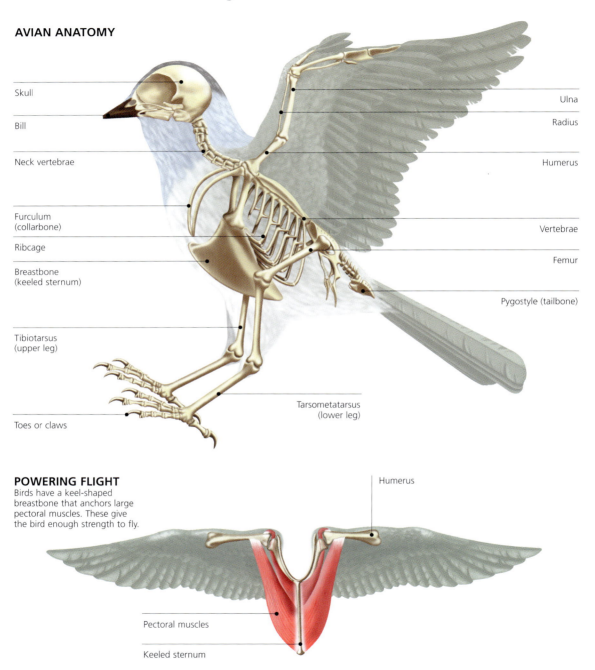

- Skull
- Bill
- Neck vertebrae
- Furculum (collarbone)
- Ribcage
- Breastbone (keeled sternum)
- Tibiotarsus (upper leg)
- Toes or claws
- Tarsometatarsus (lower leg)
- Ulna
- Radius
- Humerus
- Vertebrae
- Femur
- Pygostyle (tailbone)

POWERING FLIGHT
Birds have a keel-shaped breastbone that anchors large pectoral muscles. These give the bird enough strength to fly.

- Humerus
- Pectoral muscles
- Keeled sternum

DIGESTIVE SYSTEM

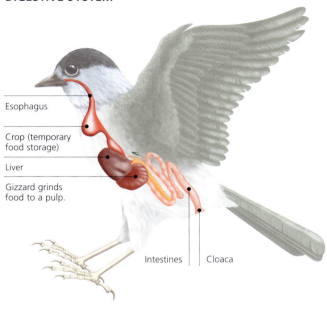

LEG MUSCLES
All birds have powerful leg muscles in the top of the leg, near their center of gravity. These are connected to the toes by long tendons.

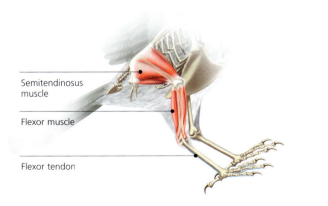

HOLLOW BONES
Most bones in a bird's body are thin-walled and hollow, with struts and braces providing maximum strength for minimum weight.

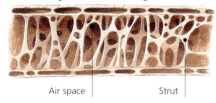

LUNG CROSS-SECTION

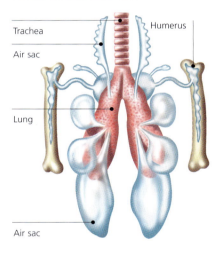

HEART CROSS-SECTION

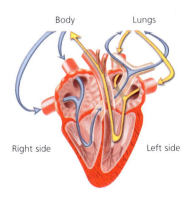

Flightless Birds

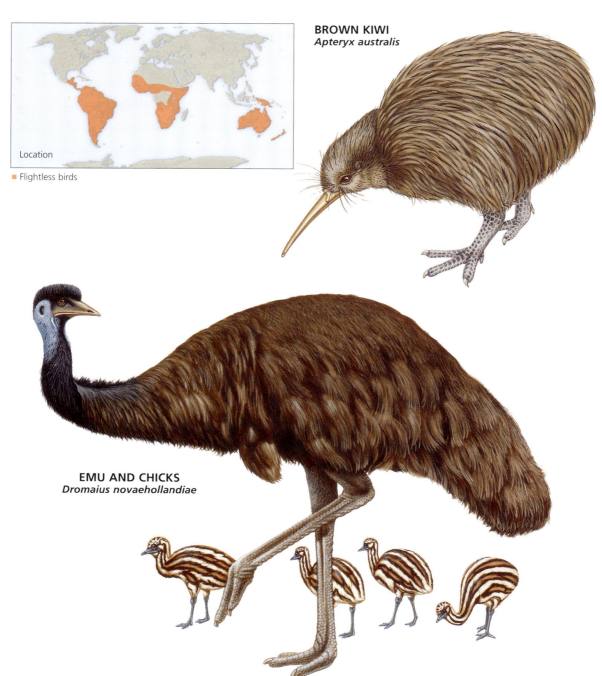

Location
- Flightless birds

BROWN KIWI
Apteryx australis

EMU AND CHICKS
Dromaius novaehollandiae

Female great tinamou head

Emu at rest

Emu egg

Great tinamou

Greater rhea

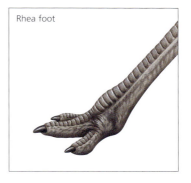
Rhea foot

Kiwi using claws to scratch for food

Ostrich tail feather

Ostrich

Kiwi head and bill

Cassowary foot

Southern cassowary

Penguins

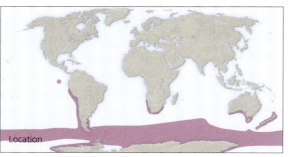

Location
■ Penguins

EMPEROR PENGUIN ANATOMY

Feathers **Foot**

Scaly, oily tips | Bend at the base | Fluffy down | Ankle feathers | Long toenails to grip the ice

EMPEROR PENGUINS SWIMMING
Emperor penguins are swift, agile swimmers, and can swim up to 20 miles (32 km) per hour. They dive through a hole in the ice and use their stiff, flipperlike wings to chase their prey.

FJORDLAND PENGUIN
Eudyptes pachyrhynchus

ROCKHOPPER PENGUIN
Eudyptes chrysocome

MAGELLANIC PENGUIN
Spheniscus magellanicus

FAIRY PENGUIN
Eudyptula minor

EMPEROR PENGUIN
Aptenodytes forsteri

CHINSTRAP PENGUIN
Pygoscelis antarctica

YELLOW-EYED PENGUIN
Megadyptes antipodes

Grebes and Divers

Location
- Grebes and dabchicks
- Divers (loons)
- Grebes, dabchicks and divers

GREAT CRESTED GREBE
Podiceps cristatus

RED-NECKED GREBE
Podiceps grisegena

PIED-BILLED GREBE WITH CHICKS
Podilymbus podiceps

Great northern

Red-throated

Black-throated

GROUP OF DIVERS (LOONS)
Gavia sp.

COMMON LOON
Gavia immer

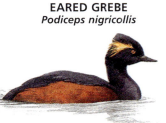

EARED GREBE
Podiceps nigricollis

HORNED GREBE
Podiceps auritus

WESTERN GREBES 'RUSHING'
Rushing involves two or more grebes running upright across the water as part of a ritualistic courtship display.

Lakeside vegetation

Wings are drawn back and flexed.

Necks are arched with head slightly bowed.

Males use this display to defend their territory.

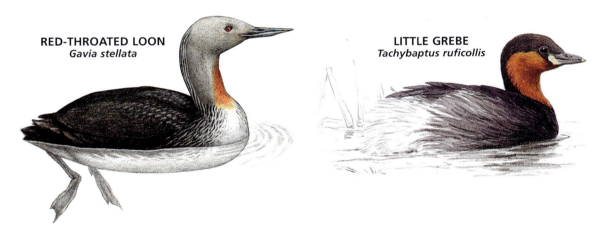

RED-THROATED LOON
Gavia stellata

LITTLE GREBE
Tachybaptus ruficollis

Pacific (left) and common loon heads

Western (top) and Clark's grebe heads

Yellow-billed loon head and neck

Albatrosses and Petrels

Location
■ Albatrosses, petrels and shearwaters

GRAY-HEADED ALBATROSS
Thalassarche chrysostoma

SHEARWATER
Puffinus sp.

CAPE PETREL
Daption capense

ANTARCTIC GIANT PETREL
Macronectes giganteus

Royal albatross mating ritual

Wandering albatross in flight

Albatross chick in nest

SHORT-TAILED SHEARWATER MIGRATION

June–August
Shearwaters spend the non-breeding season in the North Pacific Ocean.

April–May
Chick reaches adult weight and grows flight feathers, ready to leave the nest.

February–March
The young chick is fed every three days as its parents alternate between care and hunting.

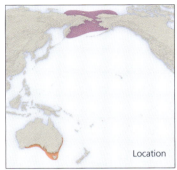

Location

■ Breeding ■ Wintering

Migratory path

September
It takes several weeks for the birds migrating south to reach their nesting colonies, along the coast of Australia.

October–November
After courting, shearwaters establish pair bonds for the year, and repair their nest site.

December–January
Both parents incubate the egg, laid late November, until it hatches in January.

251

Pelicans

Location
- Pelicans

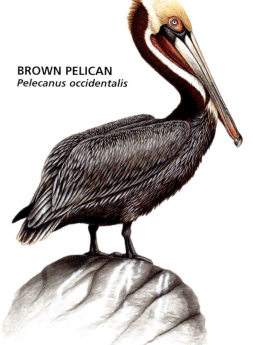
BROWN PELICAN
Pelecanus occidentalis

Shag with summer plumage

British cormorant spreading wings

Blue-footed booby on rock

Magnificent frigatebird with throat sac

Gannet at rest

Australian pelican

Herons

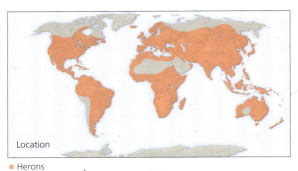
Location
- Herons

Juvenile Male Female

EURASIAN SPOONBILLS
Platalea leucorodia

Adult Juvenile

BLACK-CROWNED NIGHT HERONS
Nycticorax nycticorax

GREAT BITTERN
Botaurus stellaris

SHOEBILL WITH YOUNG
Balaeniceps rex

LITTLE EGRET
Egretta garzetta

GREAT WHITE EGRET
Egretta alba

Yellow-crowned night heron head

Roseate spoonbill spreading wings

Roseate spoonbill at rest

Juvenile little blue heron

Great blue heron in flight

African spoonbill head

Black stork wading

Great blue heron head

Scarlet ibis and skeleton

White stork in flight

Great bittern camouflaged in reeds

Great (left) and little bitterns in flight

Herons

GREEN HERONS FISHING
The green herons of Florida, U.S.A., have learnt that dropping breadcrumbs into the water attracts fish. The next generation are then taught the same tactic.

GRAY HERON
Ardea cinerea

STRIATED HERON
Butorides striatus

GREEN HERON
Butorides virescens

WALDRAPP
Geronticus eremita

LITTLE BLUE HERON
Egretta caerulea

Flamingos

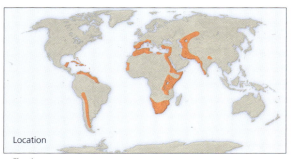

Location
- Flamingos

LESSER FLAMINGO FEEDING

Long, flexible neck bends to the water while the flamingo remains upright.

Upper jaw contains a row of slits, called lamellae, that act like a sieve.

Tongue and lower jaw strain food from the water through the lamellae.

LESSER FLAMINGO
Phoenicopterus minor

PUNA FLAMINGO
Phoenicopterus jamesi

Flamingo at rest

Flamingo feeding

Greater flamingos

GREATER FLAMINGO COLONY

Flamingos fly with their neck stretched out and long legs trailing behind.

Flamingos are highly social birds that live and breed in large lakeside colonies.

Their dramatic pink coloring is thought to come from certain algae that they eat.

A hooked bill scoops up water containing tiny animals and algae.

Waterfowl

Location
■ Waterfowl

COMMON MOORHENS
Gallinula chloropus

BARNACLE GEESE MIGRATING
Branta leucopsis
Barnacle geese fly in a characteristic v-shaped formation when migrating to and from Greenland and Europe.

Lead bird flies at the head of the v-shape, breaking up air currents for those that follow.

Migrating bircs spend a short time as lead bird, before dropping back into the flock.

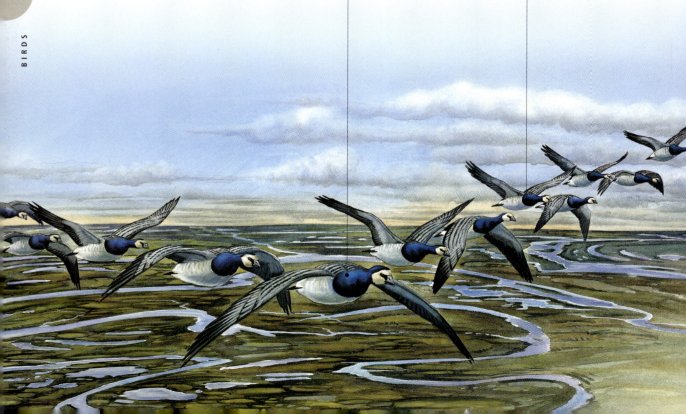

Canada goose with goslings

Velvet scoters

Gaggle of Perigord geese

Mallard foot showing webbed toes

Mallard in flight

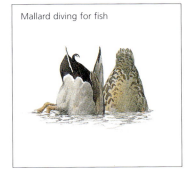
Mallard diving for fish

Ruddy duck diving

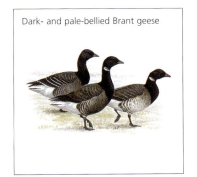
Dark- and pale-bellied Brant geese

Black swan swimming

Northen shoveler with spatulate bill

Harlequin ducks swimming

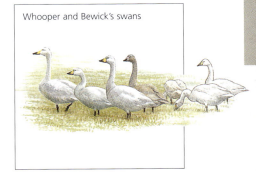
Whooper and Bewick's swans

Waterfowl

RED-BREASTED GOOSE
Branta ruficollis

KING EIDER
Somateria spectabilis

WOOD DUCK
Aix sponsa

NORTHERN SCREAMER
Chauna chavaria

GRAYLAG GEESE
Anser anser

Birds of Prey

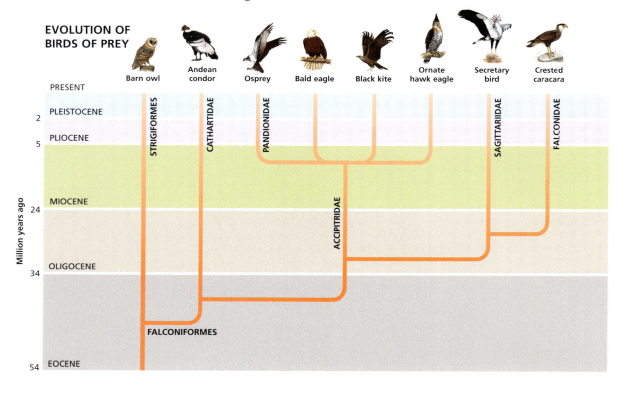

EVOLUTION OF BIRDS OF PREY

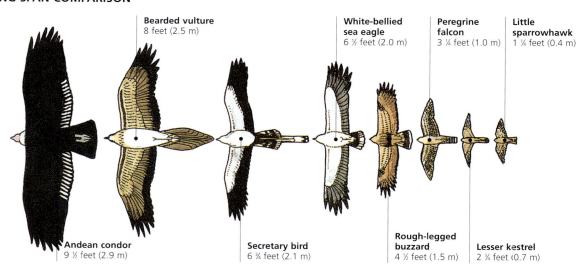

WING SPAN COMPARISON

RAPTOR ANATOMY

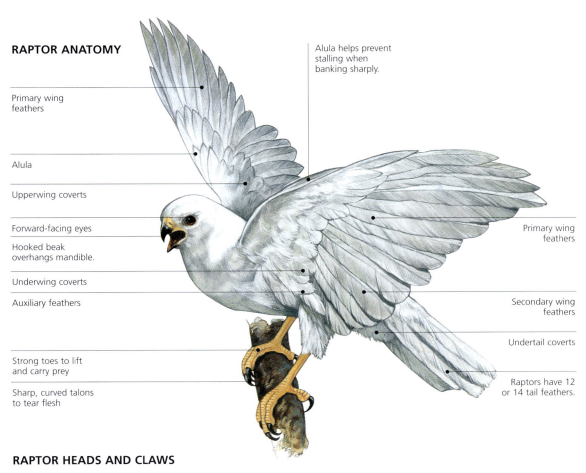

- Primary wing feathers
- Alula
- Upperwing coverts
- Forward-facing eyes
- Hooked beak overhangs mandible.
- Underwing coverts
- Auxiliary feathers
- Strong toes to lift and carry prey
- Sharp, curved talons to tear flesh
- Alula helps prevent stalling when banking sharply.
- Primary wing feathers
- Secondary wing feathers
- Undertail coverts
- Raptors have 12 or 14 tail feathers.

RAPTOR HEADS AND CLAWS

Osprey

White-backed vulture

Eurasian sparrowhawk

Harpy eagle

Osprey: fish eater

Vuture: carrion eater

Sparrowhawk: bird eater

Harpy eagle: mammal eater

Eagles, Kites and Hawks

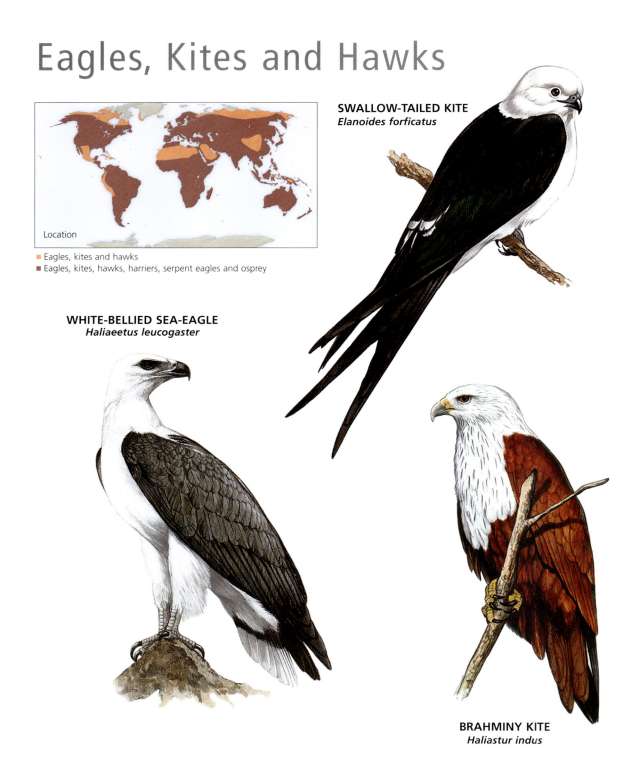

- Eagles, kites and hawks
- Eagles, kites, hawks, harriers, serpent eagles and osprey

SWALLOW-TAILED KITE
Elanoides forficatus

WHITE-BELLIED SEA-EAGLE
Haliaeetus leucogaster

BRAHMINY KITE
Haliastur indus

HOOK-BILLED KITE
Chondrohierax uncinatus

SNAIL KITE
Rostrhamus sociabilis

CRESTED SERPENT-EAGLE
Spilornis cheela

BLACK-SHOULDERED KITE
Elanus caeruleus

GREAT PHILIPPINE EAGLE
Pithecophaga jefferyi

Eagles, Kites and Hawks

PEARL KITE
Gampsonyx swainsonii

SLATE-COLORED HAWK
Leucopternis schistacea

ROUGH-LEGGED HAWK
Buteo lagopus

OSPREY
Pandion haliaetus

Falcons

Location
- Falcons and relatives
- Falcons, relatives and caracaras

AMERICAN KESTREL
Falco sparverius

GREATER KESTREL
Falco rupicoloides

STRIATED CARACARA
Phalcoboenus australis

SPOT-WINGED FALCONET
Spiziapteryx circumcinctus

Lined forest falcon

Peregrine falcon hunting pigeon

American kestrel displaying plumage

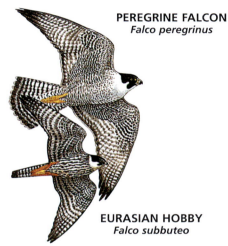

PEREGRINE FALCON
Falco peregrinus

EURASIAN HOBBY
Falco subbuteo

PYGMY FALCON
Polihierax semitorquatus

CRESTED CARACARA
Caracara plancus

PEREGRINE FALCONS MATING
The male falcon alights on the female's lowered back, and balances by flapping furiously. He lowers his tail under hers so that their cloacas meet and sperm is exchanged.

Secretary Bird and Vultures

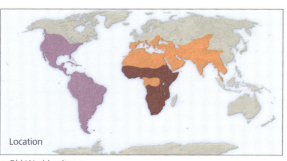

Location
- Old World vultures
- New World vultures
- Old World vultures and secretary bird

ANDEAN CONDOR
Vultur gryphus

PALM-NUT VULTURE
Gypohierax angolensis

SECRETARY BIRD
Sagittarius serpentarius

LAMMERGEIER
Gypaetus barbatus

Lappet-faced vulture spreading wings

Griffon vulture on rock

Secretary bird grabbing snake with feet

Lappet-faced vulture

Juvenile lammergeier

Black vulture head

Eurasian griffon

Juvenile (left) and adult turkey vultures

Turkey vulture in flight

273

Gamebirds

Location
■ Gamebirds

MALEO
Macrocephalon maleo

MALLEEFOWL NEST

Sand layer makes the nest warmer or cooler.

Leaf litter layer slowly rots to generate heat.

Eggs incubate at the constant temperature of about 91° F (33° C).

Malleefowls eat herbs in winter, fruits and seeds in summer.

They do not drink water, obtaining all necessary liquids from their solid diet.

Male tests temperature inside the nest with his mouth and tongue.

Chicken egg 	Wild turkey displaying plumage 	Baby fowl and egg 
Gray partridges preparing to dust bath 	Ptarmigan with winter plumage 	Ptarmigan with summer plumage 
Ruffed grouse displaying tail feathers 	Ring-necked pheasant skeleton 	Ring-necked pheasant 
Temminck's tragopan 	Satyr tragopan 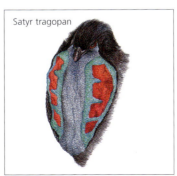	Capercaillie displaying tail feathers 

Gamebirds

INDIAN PEAFOWL
Pavo cristatus

HOATZIN
Opisthocomus hoazin

WHITE-CRESTED GUAN
Penelope pileata

GREATER PRAIRIE-CHICKEN
Tympanuchus cupido

CALIFORNIA QUAIL
Callipepla californica

REEVES' PHEASANT
Syrmaticus reevesii

VULTURINE GUINEAFOWL
Acryllium vulturinum

GAMBEL'S QUAILS
Callipepla gambelii

SPRUCE GROUSE
Falcipennis canadensis

RUFFED GROUSE
Bonasa umbellus

Cranes

Location
 Cranes

LIMPKIN
Aramus guarauna

RED-NECKED CRAKE
Rallina tricolor

WOODFORD'S RAIL
Nesoclopeus woodfordi

SIBERIAN CRANE MATING DANCE

| Male initiates mating dance. | He performs rapid bows before stretching his wings and calling. | The female responds, and is imitated. |

Cranes

TAKAHE
Porphyrio mantelli

COMMON MOORHEN
Gallinula chloropus

BUFF-BANDED RAIL
Gallirallus philippensis

WEKA
Gallirallus australis

EURASIAN COOT
Fulica atra

Juvenile (left) and adult moorhens

Kagu

Sandhill cranes migrating

Water rail chicks

Water rail

Juvenile rail skeleton

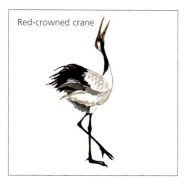
Red-crowned crane

White-quilled bustard

Plains-wanderer

Okinawa rail

Common cranes displaying

Houbara bustard

Waders and Shorebirds

WADER AND SHOREBIRD BEHAVIOR

Oystercatchers
Individual oystercatchers specialize in catching specific foods, such as worms, crabs, oysters or mussels.

Sandwich terns
Male and female sandwich terns share small fishes as part of their courtship display during the breeding season.

Horned puffins feeding
Horned puffins dive deeply into the ocean to catch small fishes, which are held in their mouth by backward-facing grooves until they are ready to eat them.

Avocet with chicks
Avocets feed by sweeping their long, upturned bill through the mud while slowly walking forward.

Turnstones
As their name suggests, turnstones walk purposefully over the mudflats, turning over stones, seaweed or wood with their beaks in search of small crustaceans and worms.

Spotted sandpiper in summer plumage

Willets in winter (left) and breeding plumage

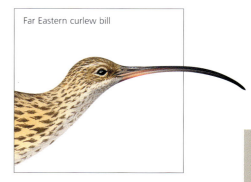
Far Eastern curlew bill

Arctic tern in flight

Jacana foot

Skua spreading wings

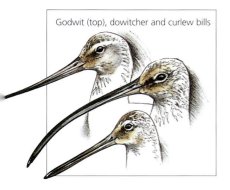
Godwit (top), dowitcher and curlew bills

Common murre egg

Golden plovers in flight

Black-headed gull facial plumage
Winter
Summer

Ruddy turnstone

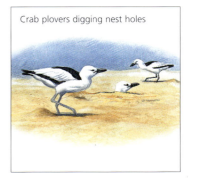
Crab plovers digging nest holes

Waders

Location
■ Waders

COMMON SNIPE
Gallinago gallinago

BANDED LAPWING
Vanellus tricolor

BLACK-BELLIED PLOVER
Pluvialis squatarola

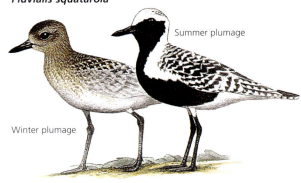
Summer plumage

Winter plumage

KILLDEER
Charadrius vociferus
The killdeer acts out a 'broken wing' display to distract and draw predators from the nest.

Winter plumage

SPOTTED SANDPIPER
Actitis macularia

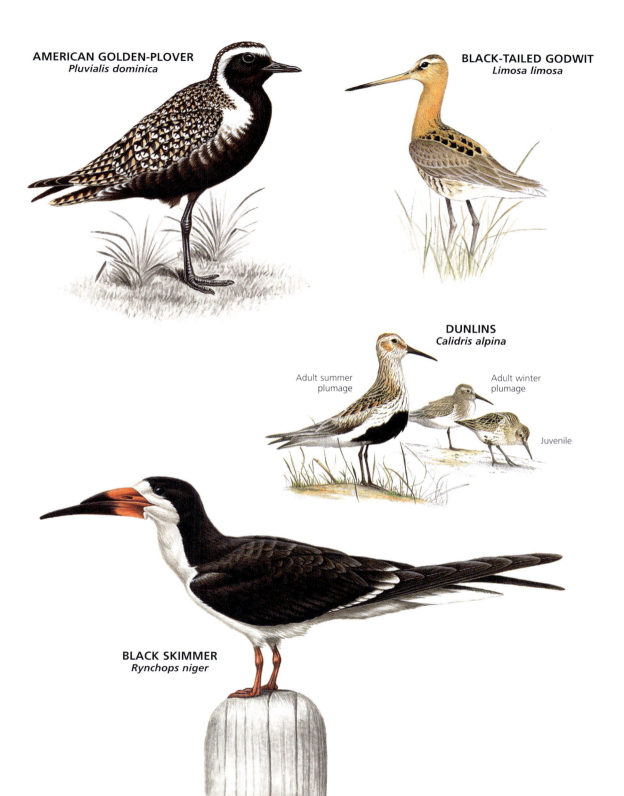

Shorebirds and Seabirds

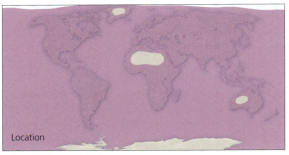

Location
■ Shorebirds and seabirds

GULL WING CUTAWAY

Wing thicker at front than back

SHOREBIRD HABITAT

Little tern | Northern gannet | Pied avocet | Yellow-legged gull | Cory's shearwater (pelicaniform) | Great cormorant

Little tern | Eurasian curlew | Black-bellied plover

Pigeons and Sandgrouse

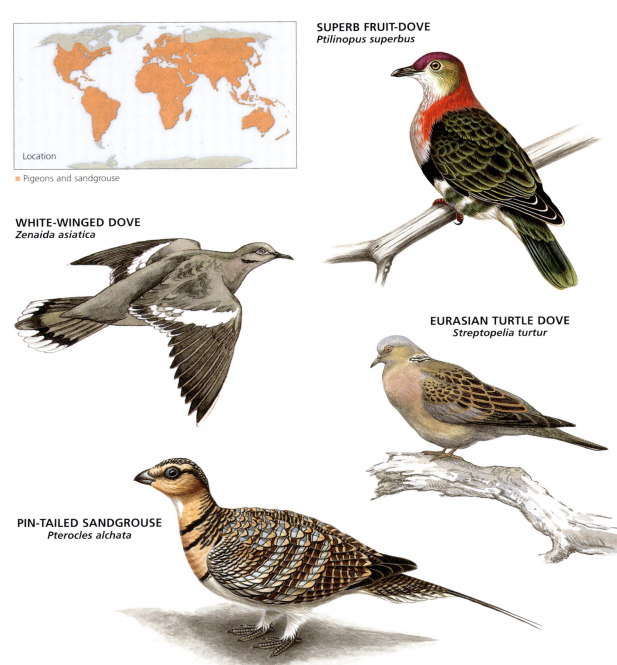

Location
■ Pigeons and sandgrouse

SUPERB FRUIT-DOVE
Ptilinopus superbus

WHITE-WINGED DOVE
Zenaida asiatica

EURASIAN TURTLE DOVE
Streptopelia turtur

PIN-TAILED SANDGROUSE
Pterocles alchata

ROCK DOVES
Columba livia

NICOBAR PIGEON
Caloenas nicobarica

Lichtenstein's sandgrouse

Male and female pin-tailed sandgrouse

Rock dove face and bill

Pigeon in flight

Crowned pigeon

White-winged dove nest

Parrots

PARROT SKULL

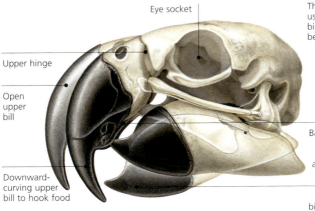

- Eye socket
- Upper hinge
- Open upper bill
- Downward-curving upper bill to hook food
- Base of upper bill used to crack seeds against curve of lower bill
- Open lower bill (mandible)

PALM COCKATOO HEAD AND BILL

The palm cockatoo uses its broad, curved bill to crush fruit, berries and seeds.

TYPES OF PARROT FEATHERS

Macaw body feather

Red-fan parrot facial feathers

Cockatiel flight feather

SCARLET MACAWS IN FLIGHT

Scarlet macaws (*Ara macao*) live in the rainforest canopies of Central and South America. Their brilliant, primary-colored plumage has made them popular with poachers and their numbers are now seriously depleted.

Australian night parrot

Kakapo

Black cockatoo

Kea

Scarlet macaw

Sulfur-crested cockatoo

Eastern rosella in flight

Gray parrot

Black-capped lory drinking nectar

Budgerigars

Lorikeet foot

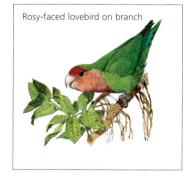
Rosy-faced lovebird on branch

Parrots

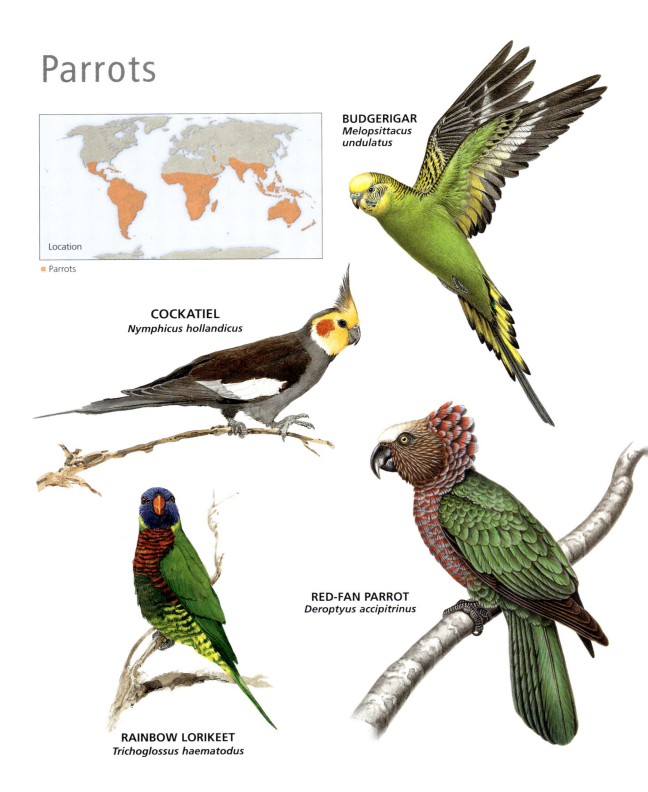

Location
■ Parrots

BUDGERIGAR
Melopsittacus undulatus

COCKATIEL
Nymphicus hollandicus

RED-FAN PARROT
Deroptyus accipitrinus

RAINBOW LORIKEET
Trichoglossus haematodus

Cuckoos and Turacos

WHITE-CHEEKED TURACO
Tauraco leucotis

GREAT BLUE TURACO
Corythaeola cristata

GREEN-BILLED MALKOHA
Phaenicophaeus tristis

AFRICAN EMERALD CUCKOO
Chrysococcyx cupreus

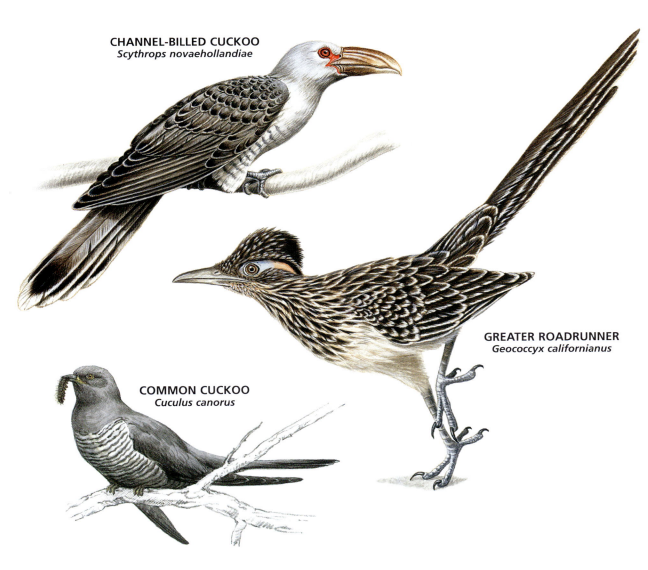

CHANNEL-BILLED CUCKOO
Scythrops novaehollandiae

GREATER ROADRUNNER
Geococcyx californianus

COMMON CUCKOO
Cuculus canorus

Cuckoo nest

Greater roadrunner in defensive posture

Violet turaco on branch

Owls

Location
■ Owls

OWL HUNTING

Glides silently toward prey
Raises wings and extends talons
Grabs prey with talons, dropping wings for balance

OWL SKULL
Some owl species have asymmetrical ears to provide greater accuracy in locating prey.

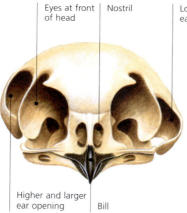
Eyes at front of head | Nostril | Lower and smaller ear opening
Higher and larger ear opening | Bill

OWL VISION
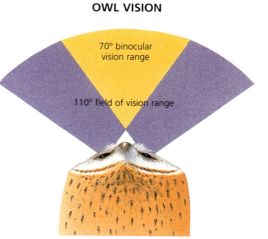
70° binocular vision range
110° field of vision range

Barn owl talons
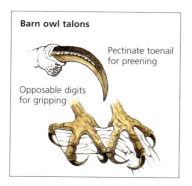
Pectinate toenail for preening
Opposable digits for gripping

Flight feathers are serrated to reduce noise.

Barn owl head and shoulders

SPOTTED OWL
Strix occidentalis

BARN OWL
Tyto alba

LITTLE OWL
Athene noctua

GREAT HORNED OWL WITH YOUNG
Bubo virginianus

Snowy owl

Barn owl chicks

Barred owl chicks in nest

Nightjars and Frogmouths

OWLETTE NIGHTJAR
Aegotheles sp.

PAURAQUE
Nyctidromus albicollis

INDIAN NIGHTJAR
Caprimulgus asiaticus

OILBIRD
Steatornis caripensis

Oilbird showing back and tail plumage

Common nighthawk catching prey

Nightjar hunting

Hummingbirds and Swifts

PURPLE-BACKED THORNBILL HOVERING
The purple-backed thornbill's wings beat rapidly in a number of directions, rotating around the flexible shoulder joint to enable the bird to hover.

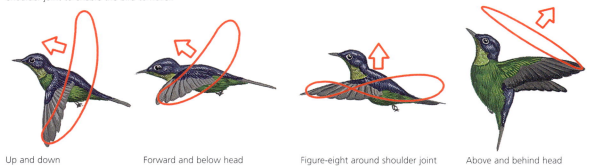

Up and down · Forward and below head · Figure-eight around shoulder joint · Above and behind head

HUMMINGBIRD FEEDING

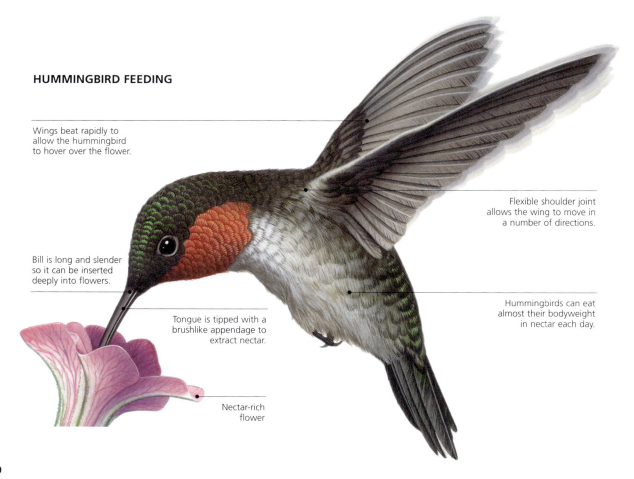

- Wings beat rapidly to allow the hummingbird to hover over the flower.
- Bill is long and slender so it can be inserted deeply into flowers.
- Tongue is tipped with a brushlike appendage to extract nectar.
- Nectar-rich flower
- Flexible shoulder joint allows the wing to move in a number of directions.
- Hummingbirds can eat almost their bodyweight in nectar each day.

Hummingbird hovering

White-throated needletail gliding

Rufous hummingbird in flight

Costa's hummingbird feeding

Swifts in flight over rooftops

Black-chinned hummingbird head

Anna's hummingbird feeding

Anna's hummingbird egg

Black-chinned hummingbird hovering

Male ruby-throated hummingbird head

Ruby-throated hummingbird on nest

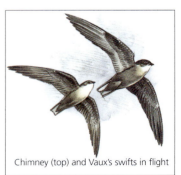
Chimney (top) and Vaux's swifts in flight

Hummingbirds and Swifts

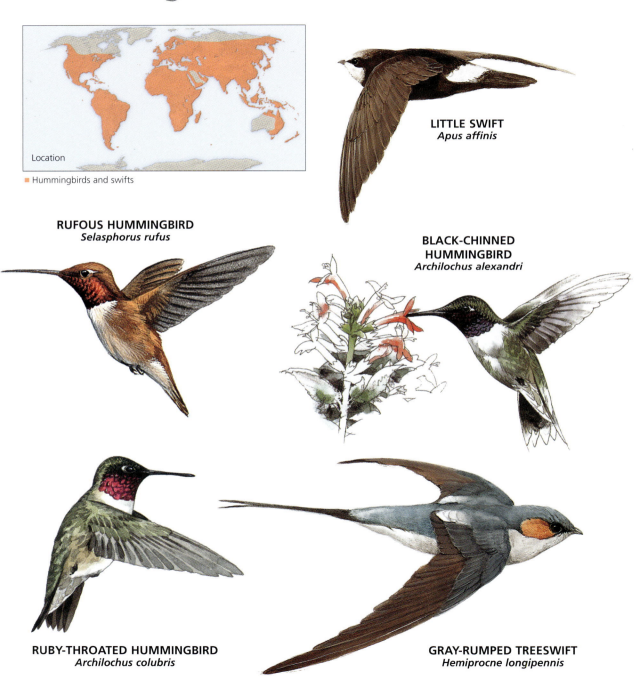

Mousebirds and Trogons

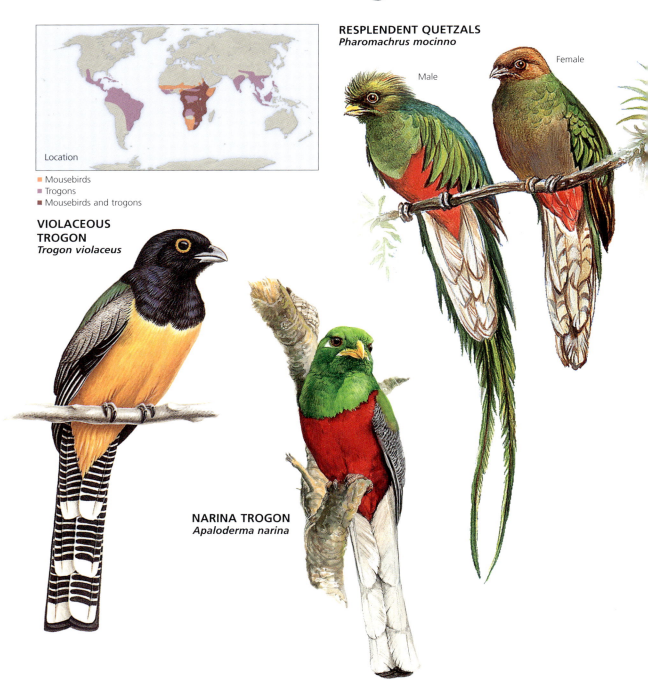

SPECKLED MOUSEBIRD
Colius striatus

RED-HEADED TROGON
Harpactes erythrocephalus

WHITE-TAILED TROGON
Trogon viridis

Scarlet-rumped trogon

Diard's trogon

Resplendent quetzal tail feathers

Kingfishers

Location

■ Kingfishers and relatives

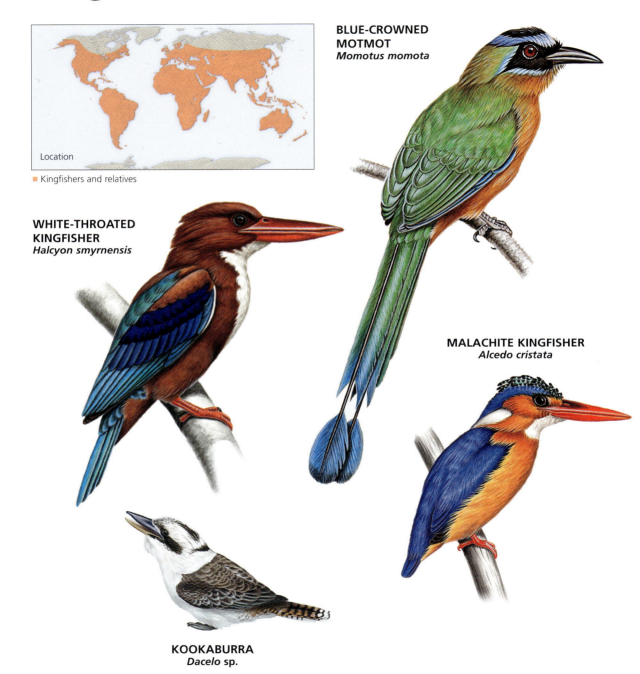

BLUE-CROWNED MOTMOT
Momotus momota

WHITE-THROATED KINGFISHER
Halcyon smyrnensis

MALACHITE KINGFISHER
Alcedo cristata

KOOKABURRA
Dacelo sp.

Cuban tody on branch

Rufous motmot

Eurasian kingfisher searching for prey

Banded kookaburra

Bee-eater in excavated nest cavity

Ground-roller on ground

Kookaburras on branch

Colorful kingfisher

Rufous-headed ground-roller

Hoopoe with young hidden in nest

Hoopoe with crest flattened

Hoopoe with crest erect

Kingfishers

NORTHERN CARMINE BEE-EATER
Merops nubicus

RAINBOW BEE-EATER
Merops ornatus

DOLLARBIRD
Eurystomus orientalis

BELTED KINGFISHER FISHING

Female flies in search of prey.

She spots a fish.

She swoops down and dives into the water.

Catching the fish in her bill, she returns to the nest.

The nest is a tunnel dug into a riverbank, above the waterline to avoid flooding.

Woodpeckers and Toucans

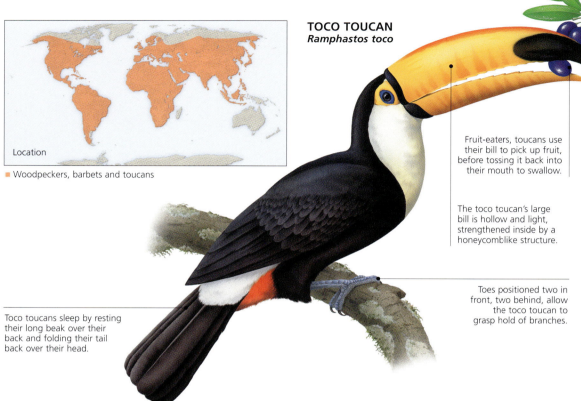

Location
■ Woodpeckers, barbets and toucans

TOCO TOUCAN
Ramphastos toco

Fruit-eaters, toucans use their bill to pick up fruit, before tossing it back into their mouth to swallow.

The toco toucan's large bill is hollow and light, strengthened inside by a honeycomblike structure.

Toes positioned two in front, two behind, allow the toco toucan to grasp hold of branches.

Toco toucans sleep by resting their long beak over their back and folding their tail back over their head.

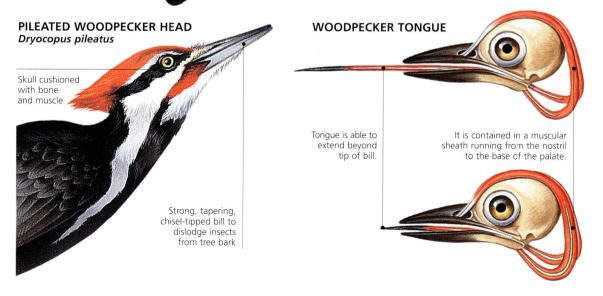

PILEATED WOODPECKER HEAD
Dryocopus pileatus

Skull cushioned with bone and muscle

Strong, tapering, chisel-tipped bill to dislodge insects from tree bark

WOODPECKER TONGUE

Tongue is able to extend beyond tip of bill.

It is contained in a muscular sheath running from the nostril to the base of the palate.

Great (left) and lesser spotted woodpeckers

Maroon woodpecker on branch

Chestnut-capped puffbird

Yellow-rumped honeyguide

Green woodpecker digging in anthill

Red-bellied woodpecker

Acorn woodpecker storing acorn

Rufous piculet

Hairy woodpecker tapping tree bark

Acorn woodpecker in flight

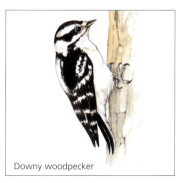
Downy woodpecker

Hairy and downy woodpecker tails

Woodpeckers and Toucans

NORTHERN FLICKER SUBSPECIES
- Yellow-shafted flicker
- Red-shafted flicker
- Gilded flicker

NORTHERN FLICKER
Colaptes auratus

KEEL-BILLED TOUCAN
Ramphastos sulfuratos

RED-AND-YELLOW BARBET
Trachyphonus erythrocephalus

Passerines

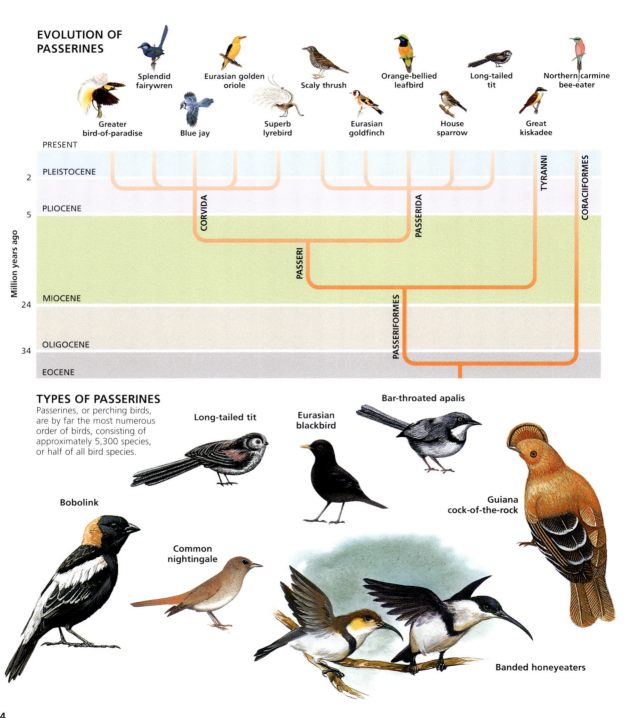

FLIGHT STAGES OF THE EUROPEAN ROBIN

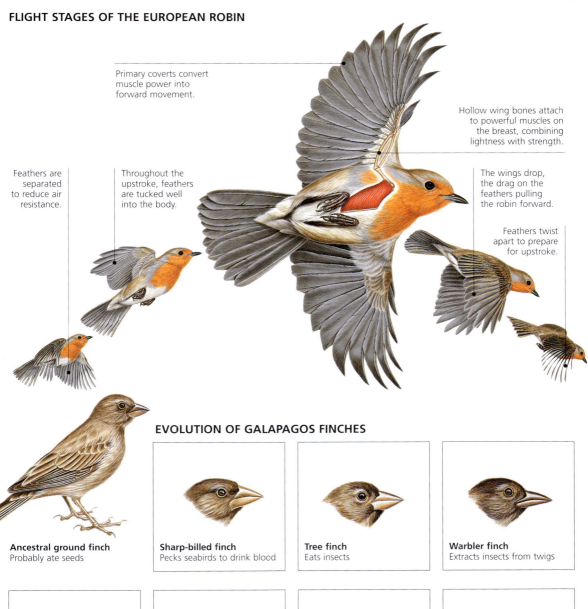

Primary coverts convert muscle power into forward movement.

Hollow wing bones attach to powerful muscles on the breast, combining lightness with strength.

Feathers are separated to reduce air resistance.

Throughout the upstroke, feathers are tucked well into the body.

The wings drop, the drag on the feathers pulling the robin forward.

Feathers twist apart to prepare for upstroke.

EVOLUTION OF GALAPAGOS FINCHES

Ancestral ground finch
Probably ate seeds

Sharp-billed finch
Pecks seabirds to drink blood

Tree finch
Eats insects

Warbler finch
Extracts insects from twigs

Woodpecker finch
Raps rotten wood for insects

Large cactus ground finch
Eats cactus flowers and seeds

Large ground finch
Cracks and eats large seeds

Tree finch
Eats plant material

Broadbills, Pittas, Ovenbirds and Tyrant Flycatchers

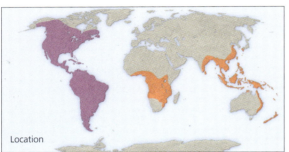

Location
- Broadbills and pittas
- Ovenbirds and tyrant flycatchers

GREEN BROADBILL
Calyptomena viridis

CASSIN'S KINGBIRD
Tyrannus vociferans

BANDED PITTA
Pitta guajana

GREAT KISKADEE
Pitangus sulphuratus

Lyrebirds, Scrub-birds, Larks, Wagtails and Swallows

Location
- Larks, wagtails and swallows
- Larks, wagtails, swallows, lyrebirds and scrub-birds

WAGTAILS
Motacilla sp.

Adult pied wagtail

Juvenile white wagtail   Adult white wagtail

WOODLARK
Lullula arborea

GRAY WAGTAIL
Motacilla cinerea

SKYLARKS FEEDING
Alauda arvensis

WATER PIPIT
Anthus spinoletta

Male (left) and female purple martins

Barn swallow

Male (left) and female swallows

Sand martins at nest cavity

Barn swallow tail feathers

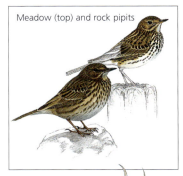
Meadow (top) and rock pipits

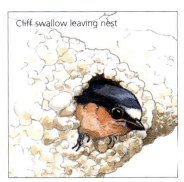
Cliff swallow leaving nest

Prairie (left) and northern horned larks

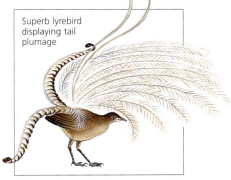
Superb lyrebird displaying tail plumage

Cliff swallow tail feathers

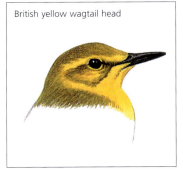
British yellow wagtail head

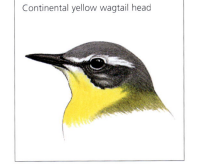
Continental yellow wagtail head

Cuckooshrikes, Bulbuls, Leafbirds, Shrikes, Vangas and Waxwings

- Cuckooshrikes, bulbuls and leafbirds
- Shrikes, vangas and waxwings
- Cuckooshrikes, bulbuls, leafbirds, shrikes, vangas and waxwings

RED-VENTED BULBUL
Pycnonotus cafer

SCARLET MINIVET
Pericrocotus flammeus

YELLOW-EYED CUCKOOSHRIKE
Coracina lineata

BOHEMIAN WAXWING
Bombycilla garrulus

Mockingbirds, Accentors, Dippers, Thrushes, Babblers and Wrens

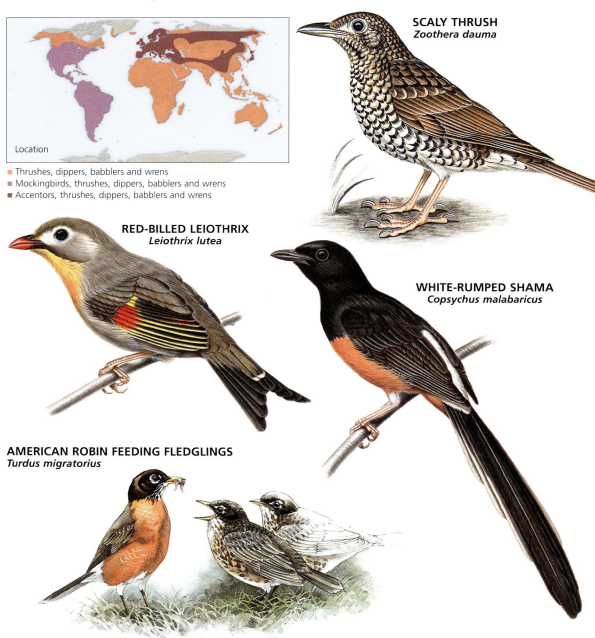

Location
- Thrushes, dippers, babblers and wrens
- Mockingbirds, thrushes, dippers, babblers and wrens
- Accentors, thrushes, dippers, babblers and wrens

SCALY THRUSH
Zoothera dauma

RED-BILLED LEIOTHRIX
Leiothrix lutea

WHITE-RUMPED SHAMA
Copsychus malabaricus

AMERICAN ROBIN FEEDING FLEDGLINGS
Turdus migratorius

Male American redstart on branch

Wren singing on branch

Wren at entrance of dome-shaped nest

Male black redstart

Canyon wren on rock

Male European robin 'redbreast'

Stonechats and whinchat (bottom)

Mockingbird in flight

Eurasian blackbird

Male western bluebird

Eastern bluebird

Dunnock

Warblers, Flycatchers, Fairywrens, Logrunners and Monarchs

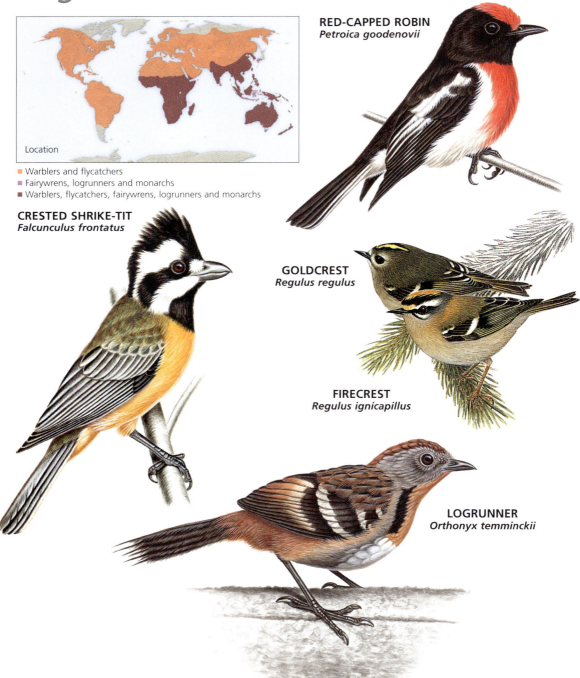

Location
- Warblers and flycatchers
- Fairywrens, logrunners and monarchs
- Warblers, flycatchers, fairywrens, logrunners and monarchs

RED-CAPPED ROBIN
Petroica goodenovii

CRESTED SHRIKE-TIT
Falcunculus frontatus

GOLDCREST
Regulus regulus

FIRECREST
Regulus ignicapillus

LOGRUNNER
Orthonyx temminckii

Tits, Nuthatches, Treecreepers and Honeyeaters

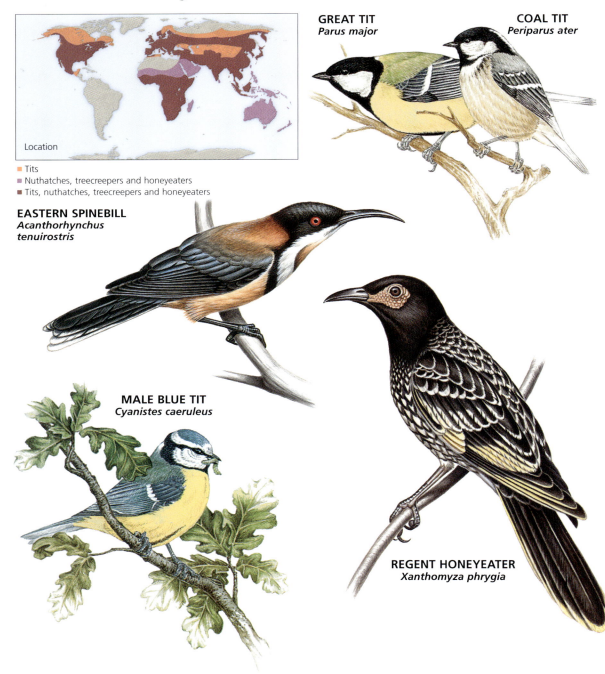

Buntings, Tanagers, Vireos and Wood Warblers

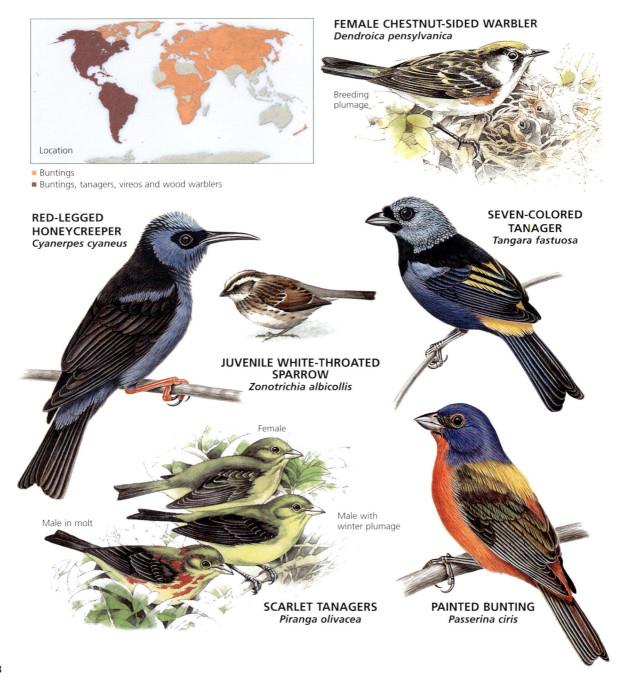

Female northern cardinal head

Male northern cardinal

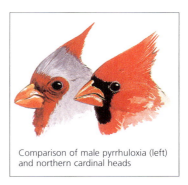
Comparison of male pyrrhuloxia (left) and northern cardinal heads

House sparrows

Slate-colored (top) and Oregon juncos

Scarlet tanager nest with eggs

Female house sparrow dustbathing

Northern parula at nest dug into moss

Adult white-striped sparrow head

Warbling (top) and Philadelphia vireos

Virginia's warbler on branch

Nashville warbler at nest entrance

Finches

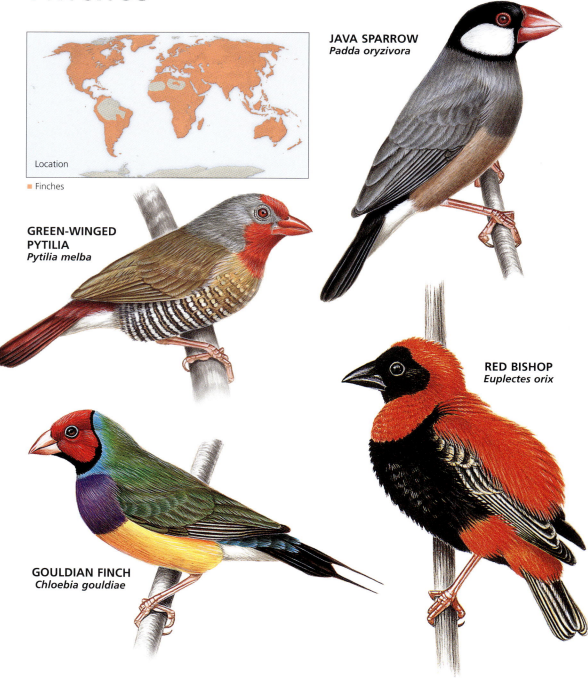

JAVA SPARROW
Padda oryzivora

GREEN-WINGED PYTILIA
Pytilia melba

RED BISHOP
Euplectes orix

GOULDIAN FINCH
Chloebia gouldiae

Eurasian bullfinch at birth
Naked skin; eyes closed

At six days
Sparse covering of down; eyes open

At 28 days
Feathers grow out of tracts along wings.

At nine months
Bullfinch fully grown

EUROPEAN GOLDFINCH
Carduelis carduelis

EUROPEAN GREENFINCHES
Carduelis chloris

Male   Female

CHAFFINCH
Fringilla coelebs

Starlings, Magpie-larks and Icterids

- Starlings
- Magpie-larks (Old World) and icterids (New World)
- Starlings, magpie-larks and icterids

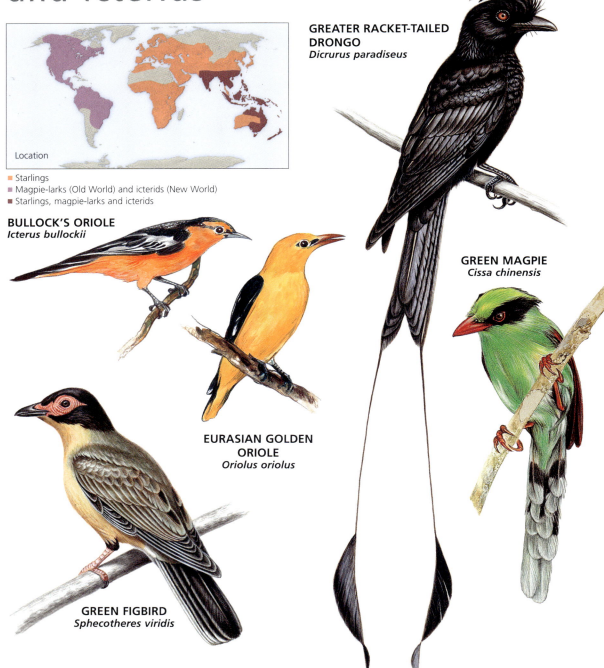

GREATER RACKET-TAILED DRONGO
Dicrurus paradiseus

BULLOCK'S ORIOLE
Icterus bullockii

GREEN MAGPIE
Cissa chinensis

EURASIAN GOLDEN ORIOLE
Oriolus oriolus

GREEN FIGBIRD
Sphecotheres viridis

Bowerbirds and Birds-of-Paradise

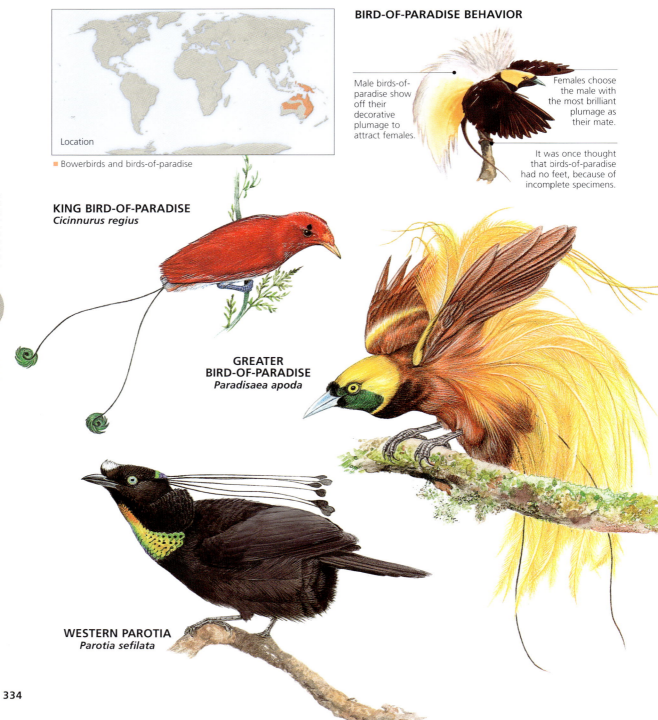

Location
■ Bowerbirds and birds-of-paradise

BIRD-OF-PARADISE BEHAVIOR

Male birds-of-paradise show off their decorative plumage to attract females.

Females choose the male with the most brilliant plumage as their mate.

It was once thought that birds-of-paradise had no feet, because of incomplete specimens.

KING BIRD-OF-PARADISE
Cicinnurus regius

GREATER BIRD-OF-PARADISE
Paradisaea apoda

WESTERN PAROTIA
Parotia sefilata

GOLDEN BOWERBIRD
Pronodura newtoniana

MACGREGOR'S BOWERBIRD
Amblyornis macgregoriae

SATIN BOWERBIRD
Ptilonorhynchus violaceus

TOOTH-BILLED BOWERBIRD
Ailuroedus dentirostris

BOWER DECORATIONS
Male bowerbirds create elaborate bowers, or courtship display areas, in one of four different types—court, mat, maypole or avenue. These are decorated with a variety of found objects, from feathers to clothes-pegs.

Moss

Twigs

Feather

Crows and Jays

Location
■ Crows and jays

RED-BILLED CHOUGH
Pyrrhocorax pyrrhocorax

AMERICAN CROW FLIGHT
Corvus brachyrhynchos

YELLOW-BILLED CHOUGH
Pyrrhocorax graculus

Crows beat their wings evenly, with strong downstrokes, to maintain a level flight path.

ROOKS
Corvus frugilegus

Juvenile

Adult

BLUE JAY
Cyanocitta cristata

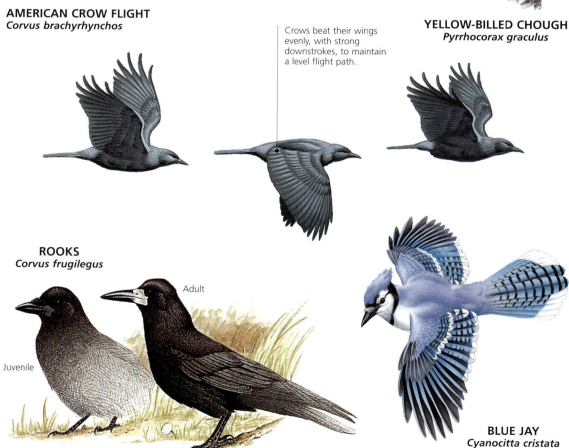

FLORIDA SCRUB-JAY
Aphelocoma coerulescens

Scrub-jays are opportunistic hunters, often stealing eggs from other songbirds' nests.

EURASIAN JACKDAW
Corvus monedula

EURASIAN JAY
Garrulus glandarius

Yellow-billed magpie head

Blue jay head showing markings

American crow (top) and common raven

Male pinyon jay on pine branch

Crested jay

Common raven head

REPTILES

INTRODUCING REPTILES

Classifying Reptiles	342
Reptile Characteristics	344
Evolution of Reptiles	346

TURTLES AND TORTOISES

Turtles and Tortoises	348
Side-necked Turtles and Land Tortoises	350
Turtles	352

CROCODILIANS

Crocodilians	354
Alligators and Caimans	356
Crocodiles and Gharials	358

AMPHISBAENIANS AND TUATARA

Amphisbaenians and Tuatara	360

LIZARDS

Lizards	362
Iguanids	364
Agamids and Chameleons	366
Iguanas and Anoles	368
Geckos	370
Beaded Lizards and Monitors	374
Skinklike Lizards	376

SNAKES

Snakes	380
Wormsnakes and Pipe Snakes	382
Boas and Pythons	384
Colubrid Snakes	386
Elapid Snakes and Vipers	390
Elapid Snakes	392
Vipers	394

Classifying Reptiles

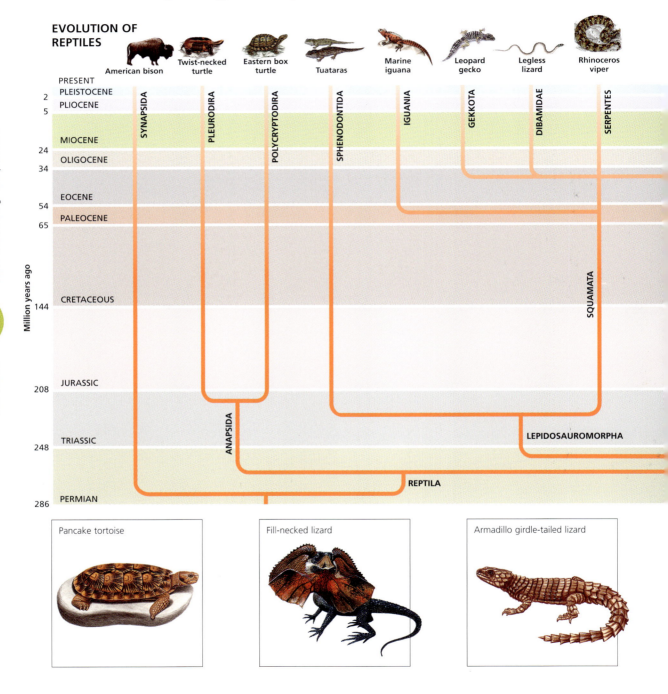

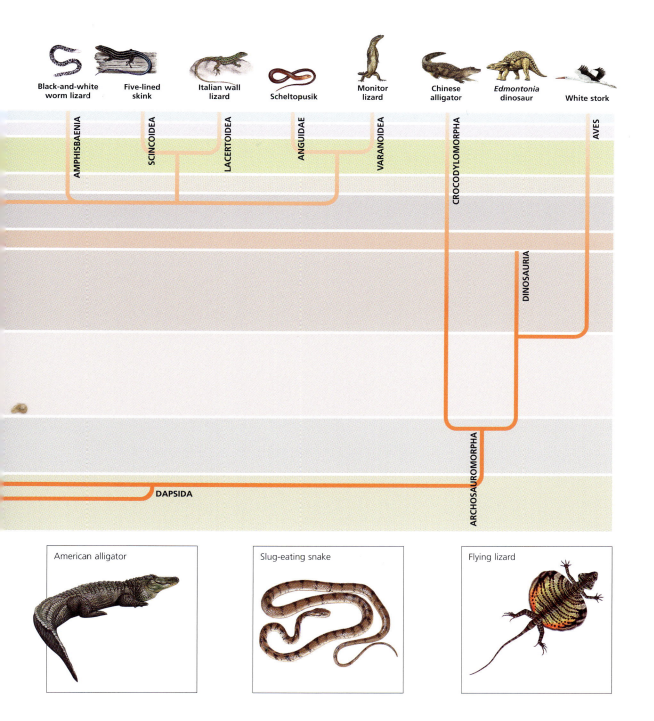

Reptile Characteristics

TYPES OF REPTILES

REPTILE SKIN

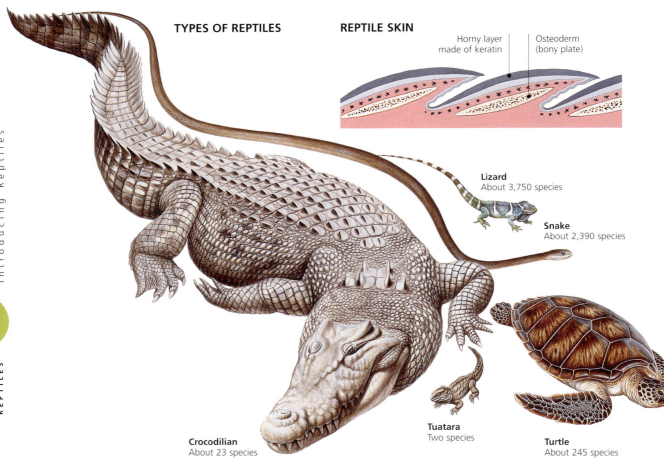

Lizard
About 3,750 species

Snake
About 2,390 species

Crocodilian
About 23 species

Tuatara
Two species

Turtle
About 245 species

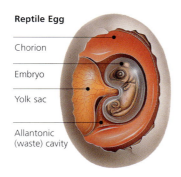

Reptile Egg
- Chorion
- Embryo
- Yolk sac
- Allantoic (waste) cavity

REPTILE SCALES

Granular scales

Keeled scales

Smooth scales

Evolution of Reptiles

EARLY REPTILES

PTERANODON
With a wingspan of 24 ½ feet (7.5 m), *Pteranodon* lived 85–75 million years ago.

DEINONYCHUS
Active dinosaur that lived 110–100 million years ago.

ICHTHYOSAURUS
Aquatic reptile that lived 206–150 million years ago.

Coelurosauravus Late Permian

Dimetrodon Permian

Hylonomus Carboniferous

Proganochelys Late Triassic

Turtles and Tortoises

LOGGERHEAD TURTLE 'CRYING'
Sea turtles, such as loggerhead turtles, 'weep' from glands around their eyes to remove salt from them. These tears are washed away by the ocean, so are visible only on land.

TORTOISE BURROW
Gopher tortoises dig burrows in the desert to escape the intensity of the heat of the day and chill of the night.

TURTLE REPRODUCTION

Mating
While some turtle species mate at sea, others mate on land. Almost all turtle species, however, lay their eggs on land.

Laying eggs
Most turtles and tortoises lay their eggs in a nest chamber, which keeps the developing eggs at a constant temperature.

Hatching out
Newly hatched turtles dig themselves out of the nest chamber. They are independent from the moment they are born.

TUCKED AWAY
Most turtles are able to draw their necks and limbs completely into their shell, protecting themselves from predators.

FROM ABOVE AND BELOW

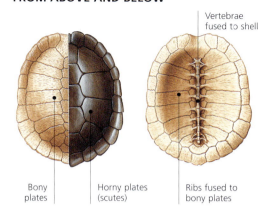

Vertebrae fused to shell

Bony plates | Horny plates (scutes) | Ribs fused to bony plates

TURTLE SHELLS

Land tortoise
Domed shell

Semi-terrestrial turtle
Flattened shell

Pond turtle
Small flattened shell

Sea turtle
Streamlined shell

TURTLE LIMBS

Land tortoise

Pond turtle

Sea turtle

Side-necked Turtles and Land Tortoises

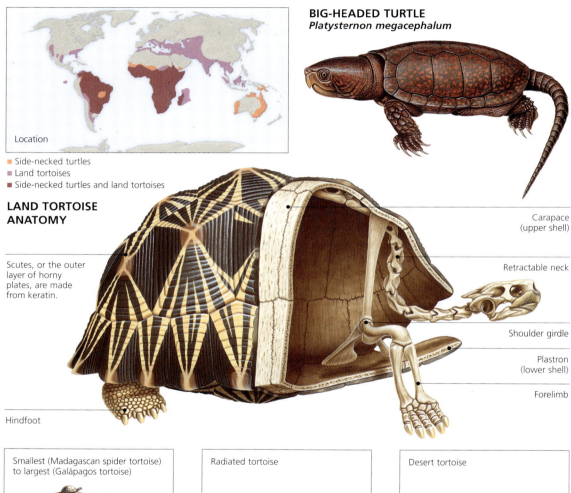

Location
- Side-necked turtles
- Land tortoises
- Side-necked turtles and land tortoises

BIG-HEADED TURTLE
Platysternon megacephalum

LAND TORTOISE ANATOMY

Scutes, or the outer layer of horny plates, are made from keratin.

Carapace (upper shell)
Retractable neck
Shoulder girdle
Plastron (lower shell)
Forelimb

Hindfoot

Smallest (Madagascan spider tortoise) to largest (Galápagos tortoise)

Radiated tortoise

Desert tortoise

YELLOW-SPOTTED AMAZON RIVER TURTLE
Podocnemis unifilis

TWIST-NECKED TURTLE
Platemys platycephala platycephala

SADDLEBACK TORTOISES
Geochelone sp.

Cactus plants form a major part of the saddleback tortoise's diet when there is no groundwater to drink.

The tortoises can stretch their necks out to reach high-growing food, or tuck them fully into their shells.

By lowering their hindlegs to the ground, saddleback tortoises can move their inflexible bodies higher.

The domed shells of many large tortoise species are honeycombed, not solid, to be less heavy.

Turtles

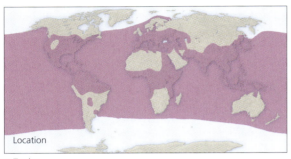
Location
■ Turtles

MALE PAINTED TURTLE
Chrysemys picta belli

MALAYAN SNAIL-EATING TURTLE
Malayemys subtrijuga

SOUTHERN LOGGERHEAD MUSK TURTLE
Sternotherus minor minor

EASTERN BOX TURTLE
Terrapene carolina carolina

SPINED TURTLE
Heosemys spinosa

Raised nostrils on snout allow turtle to breathe out of water.

Forelimbs have evolved into flippers.

Marine turtles build up large fat reserves to feed off when migrating to breed and lay eggs.

PACIFIC HAWKSBILL TURTLE
Eretmochelys imbricata bissa

EASTERN SPINY SOFTSHELL TURTLE
Apalone spinifera spinifera

SOFTSHELL TURTLES
Unlike most turtle species, softshell turtles do not have hard carapaces, but flattened shells of leathery skin. Well-adapted for life in the water, they have webbed feet and a snorkle-shaped snout.

Crocodilians

CROCODILE HATCHING

GUARDING THE NEST
Female crocodilians cover their egg nests with warm rotting plant material, and do not leave until their young are hatched.

CARRYING YOUNG
Alligator hatchlings are carried in the mother's mouth to a nearby pond. The female will protect the hatchlings from predators for several weeks.

CROCODILIAN SKULLS

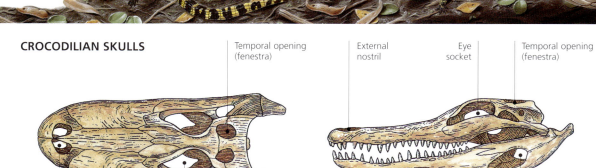

Temporal opening (fenestra) · External nostril · Eye socket · Temporal opening (fenestra)

External nostril · Eye socket · Lower jaw opening

CROCODILE GAITS

Crawling

Walking

Galloping

ANKLE COMPARISON

Crocodilian	Typical reptile
Tibia	Tibia
Fibula	Fibula
Astragalus	Astragalus

Line of hinge

SNOUT COMPARISON

American crocodile

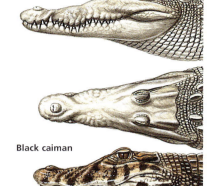

Black caiman

Gharial

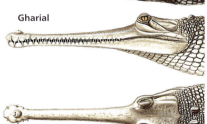

BREATHING HALF-SUBMERGED

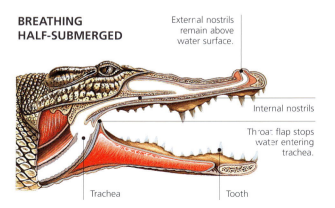

External nostrils remain above water surface.
Internal nostrils
Throat flap stops water entering trachea.
Trachea
Tooth

CROCODILE TOOTH

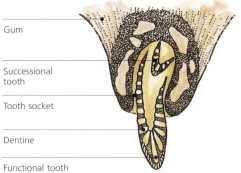

Gum
Successional tooth
Tooth socket
Dentine
Functional tooth

355

Alligators and Caimans

Location
- Alligators
- Caimans

COMMON CAIMAN
Caiman crocodilus

AMERICAN ALLIGATOR HUNTING
Alligator mississippiensis

American alligators often hunt near water-bird colonies, where they feed on the fish that gather to eat guano. The birds are not safe either—alligators can leap from the water to snatch fledglings or unwary adults from the branches.

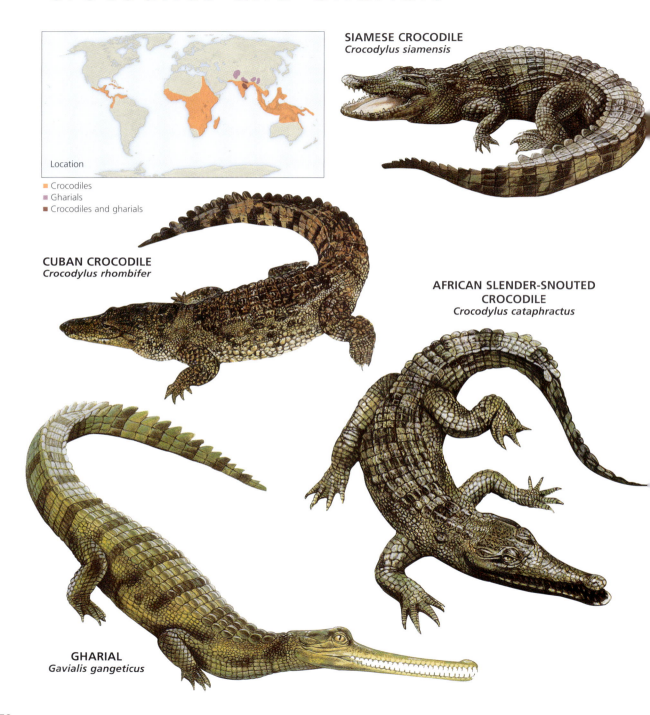

MALAYAN GHARIAL
Tomistoma schlegelii

NILE CROCODILE
Crocodylus niloticus

ORINOCO CROCODILE
Crocodylus intermedius

CROCODILE SUBMERGED
Conservative hunters, crocodiles lie submerged with only their eyes, ears and noses above the water, waiting for prey. When they strike, crocodiles move undetected through the water and lunge up onto the bank, grabbing their prey before quickly drowning, then eating it.

Amphisbaenians and Tuatara

Location
- Tuatara
- Amphisbaenians (worm-lizards)

ADAPTATIONS FOR DIGGING

Round-snouted
Push forward, then around

Spade-snouted
Push forward, then up

Chisel-snouted
Rotate in opposing directions

Keel-snouted
Push forward, then sideways

BLACK AND WHITE WORM-LIZARD
Amphisbaena fuliginosa

KEEL-SNOUTED WORM-LIZARD
Family *Amphisbaenidae*

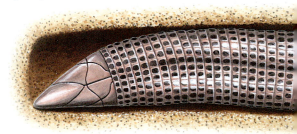

SPADE-SNOUTED WORM-LIZARD
Family *Amphisbaenidae*

TUATARA

Tuatara can be found only on about 30 small islands off the northern coast of New Zealand.

Teeth fused to jawbone

'Tuatara' means 'lightning back' in Maori, which refers to the crest of spines along the males' back.

Tuatara are not fast runners, so lie in wait for prey to approach close enough to pounce.

TUATARA
Sphenodon sp.

Female

Male

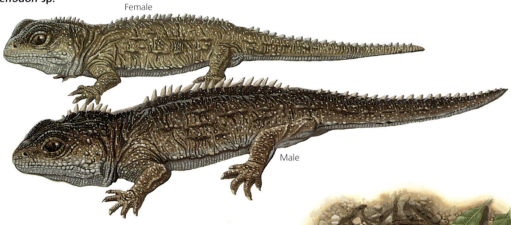

TUATARA BURROW

Burrow entrance

The nocturnal tuatara spends the day basking in the sun in front of its burrow, or asleep inside.

361

Lizards

EVOLUTION OF LIZARDS

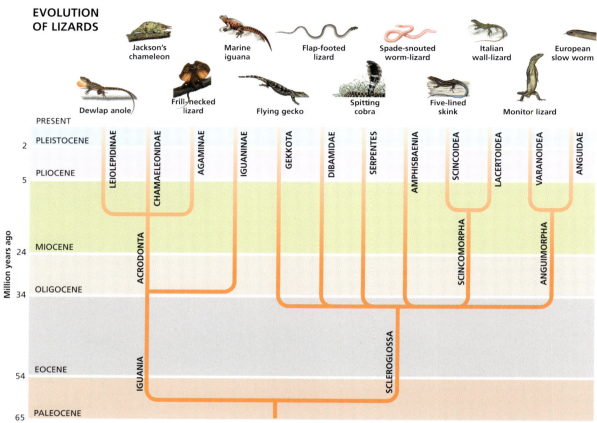

LIZARD TOES
Lizard toes have adapted to specific uses and environments, such as burrowing or climbing.

No toes

Reduced toes

Well-developed toes

FOOT SPECIALIZATIONS

Fringe-toed lizard

Fringe helps to cool feet.

Desert gecko

Webbing to grip on sand

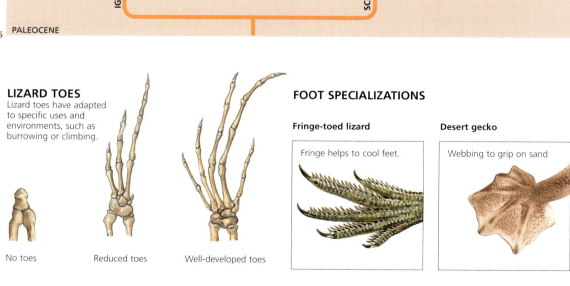

Chameleon (above): grabs prey with long, sticky, mucus-coated tongue.

LIZARD TAILS

Shingleback: fat storage

Chameleon: prehensile

Leaf-tailed gecko: camouflage

Skink: streamlined

TONGUE ADAPTATIONS

European slow worm: catches prey with pointed tongue.

Shingleback lizard: startles prey with flattened tongue.

Goanna: detects prey with forked tongue.

LIZARD EYE
Some lizards have a clear area on their eyelid to enable them to watch for predators while their eyes are closed.

Iguanids

Location
- Agamids and chameleons
- Iguanas and anoles
- Agamids, chameleons, iguanas and anoles

HELMETED IGUANA HEAD

The function of the casque, or crest, extending from head to neck, is unknown.

Like chameleons, helmeted iguanas can change color.

Pouch under the chin expands when iguana defends its territory.

COLLARED LIZARD BEHAVIOR

Crouches down — Pushes up to appear bigger — Crouches down again

Rhinoceros iguanas change color with the heat of the sun.

Thorny devil

Basilisk running on water

Regal horned lizard squirting blood

Common iguana head

Bearded dragon expanding throat

Agamids and Chameleons

CHANGING COLOR

Male chameleons flush with color to warn rival males of their territories. This species changes from camouflaging green to an angry red.

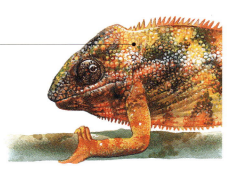

SAIL-TAILED WATER LIZARD
Hydrosaurus ambionensis

IRANIAN TOAD-HEADED AGAMA
Phrynocephalus persicus

COMMON AGAMA
Agama agama

THORNY DEVIL
Moloch horridus

Iguanas and Anoles

GREEN BASILISK
Basiliscus plumifrons

SHORT-HORNED LIZARD
Phrynosoma douglassii

DESERT SPINY LIZARD
Sceloporus magister

FIJIAN CRESTED IGUANA
Brachylophus vitiensis

COLLARED LIZARD
Crotaphytus collaris

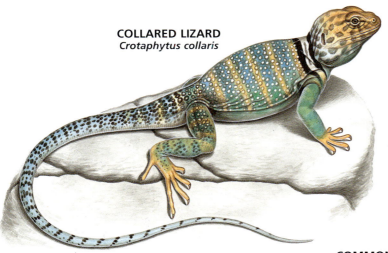

COMMON BASILISK
Basiliscus basiliscus

MARINE IGUANAS

Marine iguanas of the Galápagos Islands are the only species of lizard that depend on the sea to survive.

Male marine iguanas change color to green and red to show their readiness to mate.

The cold ocean waters mean that the iguanas spend hours basking in the sun to regain enough warmth to feed again.

Geckos

Location
- Geckos and relatives

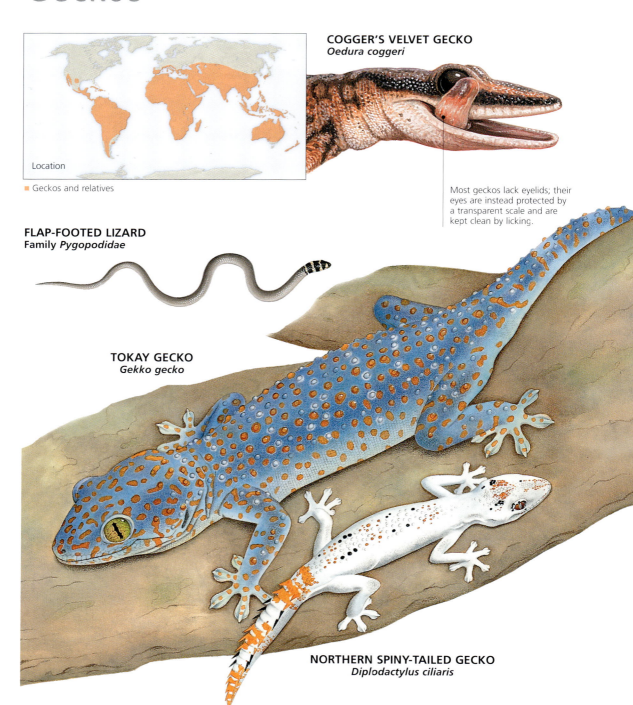

COGGER'S VELVET GECKO
Oedura coggeri

Most geckos lack eyelids; their eyes are instead protected by a transparent scale and are kept clean by licking.

FLAP-FOOTED LIZARD
Family *Pygopodidae*

TOKAY GECKO
Gekko gecko

NORTHERN SPINY-TAILED GECKO
Diplodactylus ciliaris

YELLOW-HEADED GECKO
Gonatodes albogularis fuscus

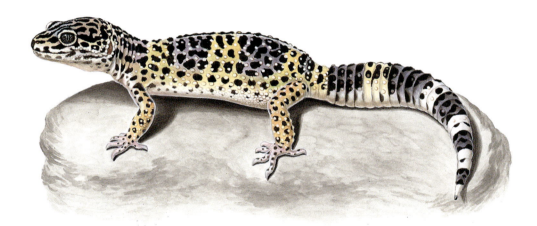
LEOPARD GECKO
Eublepharus macularius

GRAY'S BOW-FINGERED GECKO
Cyrtodactylus pulchellus

COMMON WONDER GECKO
Teratoscincus scincus

Geckos

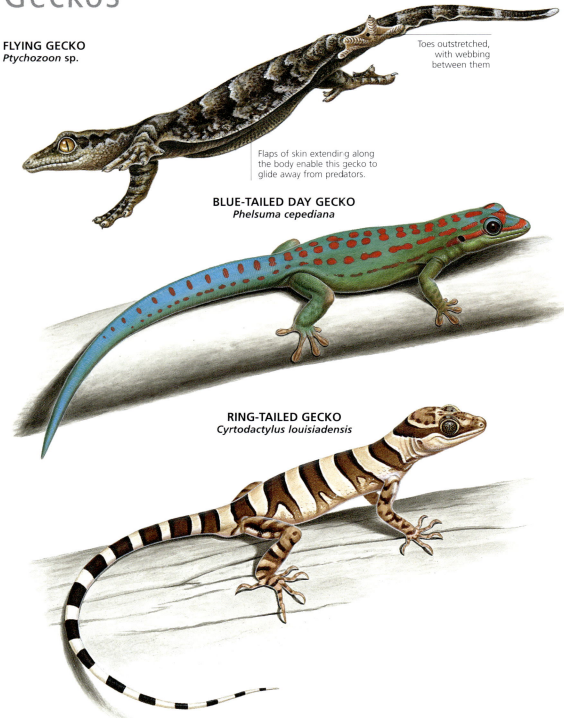

FLYING GECKO
Ptychozoon sp.

Toes outstretched, with webbing between them

Flaps of skin extending along the body enable this gecko to glide away from predators.

BLUE-TAILED DAY GECKO
Phelsuma cepediana

RING-TAILED GECKO
Cyrtodactylus louisiadensis

SOUTHERN SPOTTED VELVET GECKO
Oedura tryoni

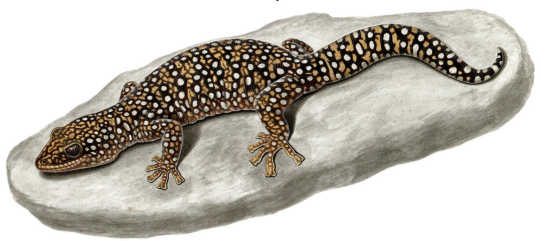

MOURNING GECKO
Lepidodactylus lugubris

Female mourning geckos are able to reproduce without fertilizing their eggs, and therefore without needing a male.

COMMON WALL GECKO
Tarentola mauritanica

The common wall gecko, or Moorish gecko, is one of only four species still found in Europe.

ISRAELI FAN-FINGERED GECKO
Ptyodactylus puiseuxi

Members of this rock-dwelling genus are found in Africa and the Middle East.

Beaded Lizards and Monitors

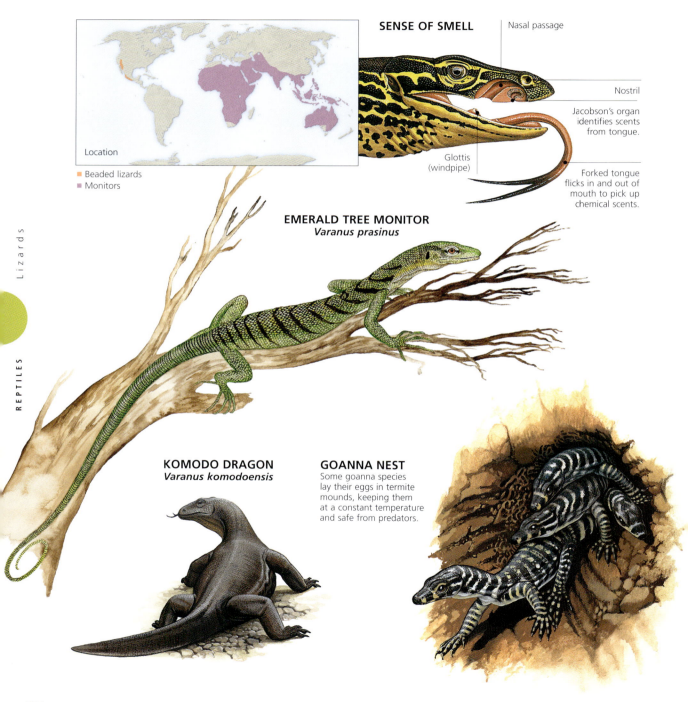

Location
- Beaded lizards
- Monitors

SENSE OF SMELL
Nasal passage
Nostril
Jacobson's organ identifies scents from tongue.
Glottis (windpipe)
Forked tongue flicks in and out of mouth to pick up chemical scents.

EMERALD TREE MONITOR
Varanus prasinus

KOMODO DRAGON
Varanus komodoensis

GOANNA NEST
Some goanna species lay their eggs in termite mounds, keeping them at a constant temperature and safe from predators.

Skinklike Lizards

Skinklike Lizards

MALE DWARF FLAT LIZARD
Platysaurus guttatus

MENORA WALL LIZARD
Podarcis perspicillata

HARIA LIZARD
Gallotia Atlantica

PYGMY BLUE-TONGUE SKINK
Tiliqua sp.

JUNGLE RUNNER
Ameiva ameiva

JUVENILE FIVE-LINED SKINK
Eumeces fasciatus

ITALIAN WALL LIZARD
Podarcis sicula

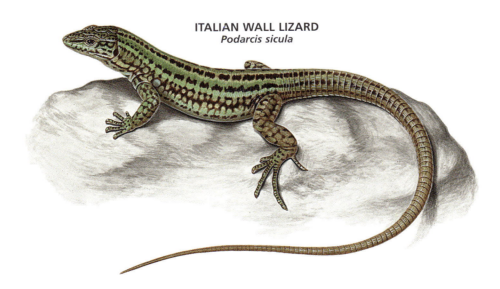

Snakes

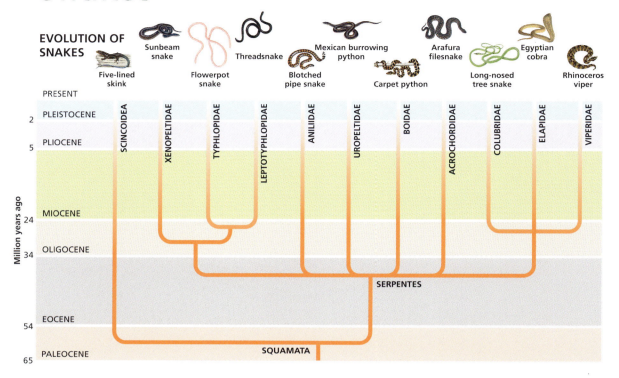

SNAKE ANATOMY

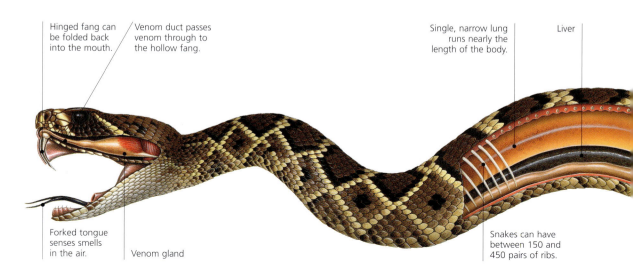

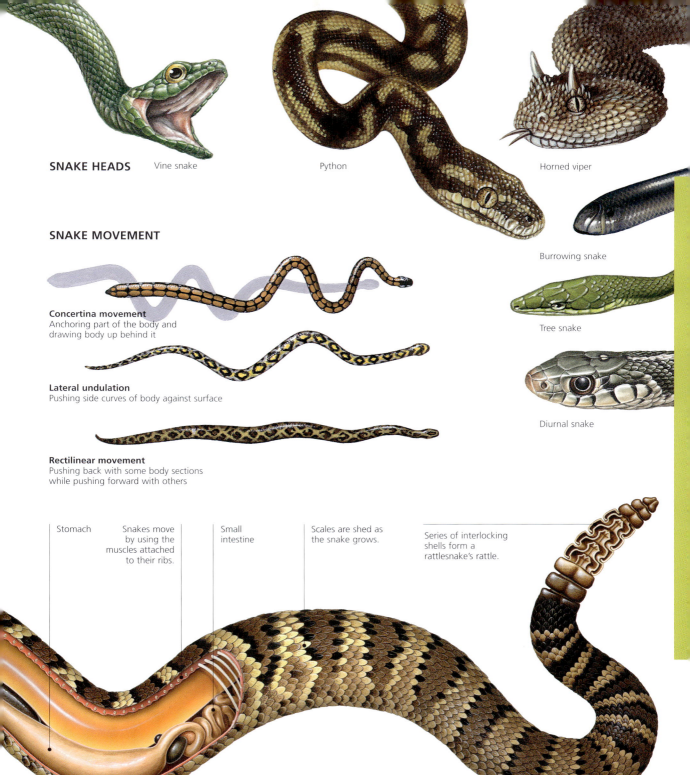

SNAKE HEADS — Vine snake, Python, Horned viper, Burrowing snake, Tree snake, Diurnal snake

SNAKE MOVEMENT

Concertina movement
Anchoring part of the body and drawing body up behind it

Lateral undulation
Pushing side curves of body against surface

Rectilinear movement
Pushing back with some body sections while pushing forward with others

Stomach

Snakes move by using the muscles attached to their ribs.

Small intestine

Scales are shed as the snake grows.

Series of interlocking shells form a rattlesnake's rattle.

Wormsnakes and Pipe Snakes

Location
- Wormsnakes
- Pipe snakes and sunbeam snake
- Wormsnakes, pipe snakes and sunbeam snake

BLIND SNAKE WITH PREY
Unlike most snakes, which eat a single large prey, blind snakes eat small prey more frequently.

Ant

Ant eggs are the main diet of blind snakes.

THREADSNAKE
Leptotyphlops humilis

FLOWERPOT SNAKE
Ramphotyphlops braminus

SUNBEAM SNAKE
Xenopeltis unicolor

MEXICAN BURROWING PYTHON
Loxocemus bicolor

SOUTH AMERICAN CORAL PIPE SNAKE
Anilius scytale

DRUMMOND HAY'S EARTH SNAKE
Rhinophis drummondhayi

BLOTCHED PIPE SNAKE
Cylindrophus maculatus

RED CYLINDER SNAKE
Cylindrophis rufus

Boas and Pythons

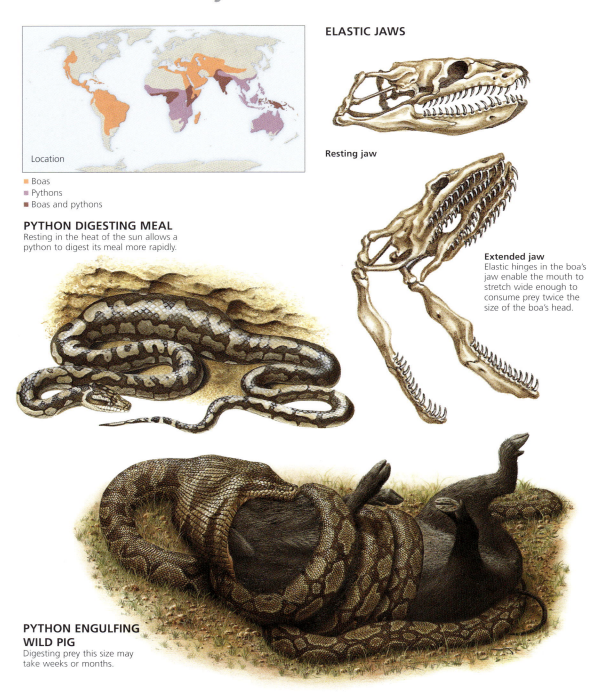

Location
- Boas
- Pythons
- Boas and pythons

ELASTIC JAWS

Resting jaw

Extended jaw
Elastic hinges in the boa's jaw enable the mouth to stretch wide enough to consume prey twice the size of the boa's head.

PYTHON DIGESTING MEAL
Resting in the heat of the sun allows a python to digest its meal more rapidly.

PYTHON ENGULFING WILD PIG
Digesting prey this size may take weeks or months.

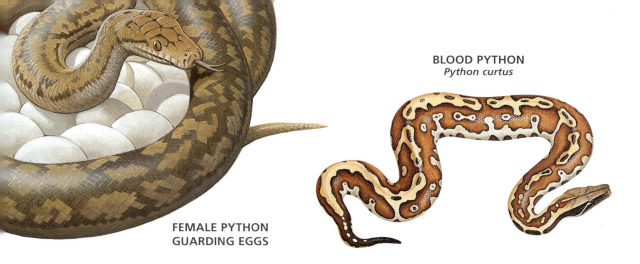

FEMALE PYTHON GUARDING EGGS

BLOOD PYTHON
Python curtus

SIZE COMPARISON

Anaconda (species of boa): 33 feet (10 m)

Boa constrictor: 14½ feet (4.5 m)

Eastern diamondback rattlesnake: 7 feet (2.2 m)

Yellow-bellied sea snake: 2½ feet (0.8 m)

BRAZILIAN RAINBOW BOA
Epicrates cenchria cenchria

CARPET PYTHON
Morelia spilota

Colubrid Snakes

Location
- Colubrid snakes
- Colubrid snakes and filesnakes

WHITE-BELLIED MANGROVE SNAKE
Fordonia leucobalia

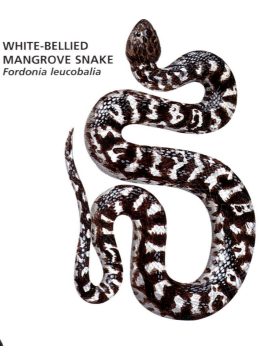

ARAFURA FILESNAKE
Acrochordus arafurae

LONG-NOSED TREE SNAKE
Ahaetulla nasuta

AFRICAN TWIG SNAKE
Thelotornis capensis

LITTLE FILESNAKE
Acrochordus granulatus

COMMON GRASS SNAKE
Natrix natrix

SPOTTED HOUSE SNAKE
Lamprophis guttatus

COMMON KING SNAKE
Lampropeltis getulus

TOAD-EATER SNAKE
Xenodon rabdocephalus

Colubrid Snakes

BLUNT-HEADED TREE SNAKE
Imantodes cenchoa

SPOTTED HARLEQUIN SNAKE
Homoroselaps lacteus

MANDARIN RATSNAKE
Elaphe mandarina

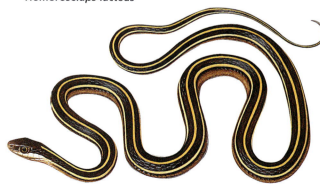

EASTERN RIBBON SNAKE
Thamnophis sauritus sauritus

GREEN VINE SNAKE
Oxybelis sp.

Even though green vine snakes can grow to 7 feet (2 m) in length, their bodies are no more than ½ inch (1.3 cm) wide.

Their slender bodies can curl easily around branches and vines as they search for small birds in nests.

Elapid Snakes and Vipers

TYPES OF FANGS

Colubrid snakes
Rear, fixed, grooved fangs

Elapid snakes
Front, fixed, hollow fangs

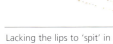

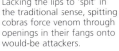

Vipers
Front, swinging, hollow fangs

RATTLESNAKE HEAD
Like vipers, rattlesnakes have swinging fangs, and can sense heat so accurately they can strike prey in total darkness.

SPITTING COBRA

Lacking the lips to 'spit' in the traditional sense, spitting cobras force venom through openings in their fangs onto would-be attackers.

The cobra's hood flap is made from skin only, which extends as the elongated ribs expand when it feels angry or threatened.

COBRA SKELETON

Elongated ribs

SIDEWINDER LOCOMOTION

The sidewinder moves across hot, loose sand by anchoring its head and tail, lifting its trunk sideways across the ground. The head and tail then join the trunk.

Tracks left in the sand dune

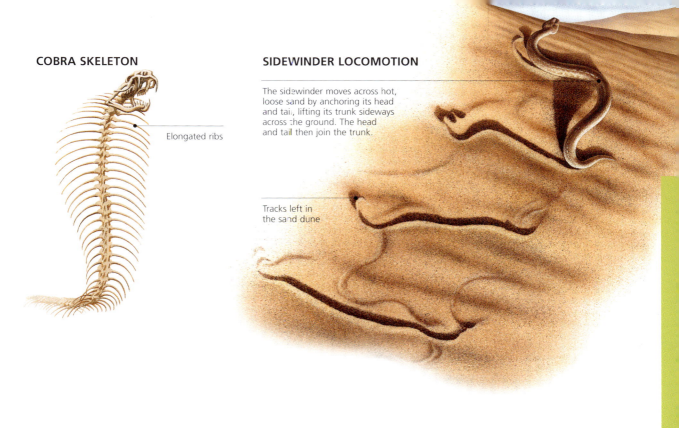

RATTLESNAKE RATTLE

Body scales

Interlocking shells

The rattle is used to distract prey, mesmerizing would-be attackers until the snake is ready to strike.

Elapid Snakes

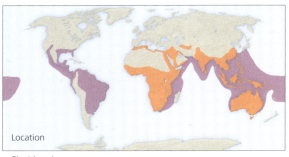

Location
- Elapid snakes
- Coral snakes and sea snakes

BLUE CORAL SNAKE HEAD
Maticora bivirgata

The venom duct of the blue coral snake extends almost a third of the way along the snake's body.

EGYPTIAN COBRA
Naja haje

YELLOW-LIPPED SEA KRAIT
Laticauda colubrina

ARIZONA CORAL SNAKE
Micruroides euryxanthus euryxanthus

Vipers

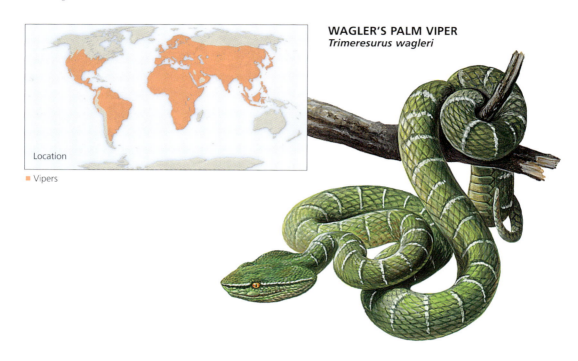

WAGLER'S PALM VIPER
Trimeresurus wagleri

Location — Vipers

URUTU
Bothrops alternatus

AMPHIBIANS

INTRODUCING AMPHIBIANS
Classifying Amphibians 400

SALAMANDERS AND NEWTS
Salamanders and Newts 402

CAECILIANS
Caecilians 408

FROGS AND TOADS
Frogs and Toads 410
Primitive Frogs 412
American and Australasian
  Frogs 414
Toads, Treefrogs and Glass
  Frogs 416
True Frogs and Poison Frogs 418
Other Frogs and Toads 420

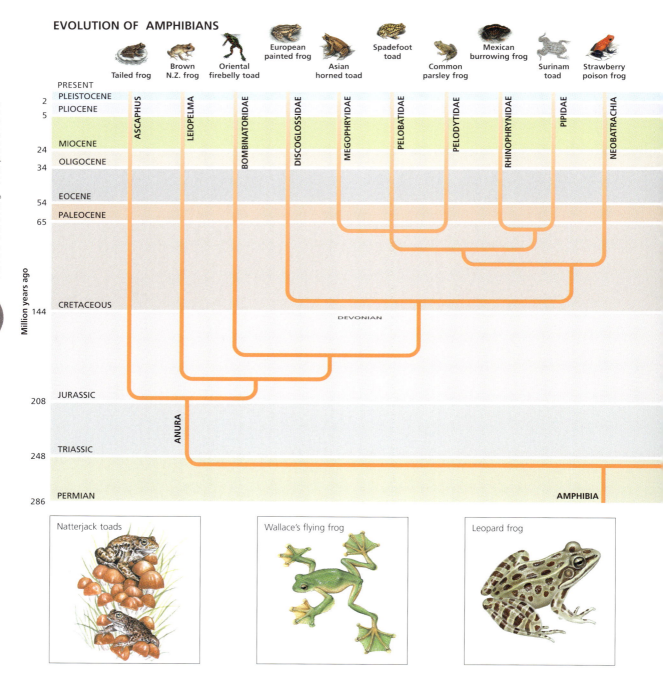

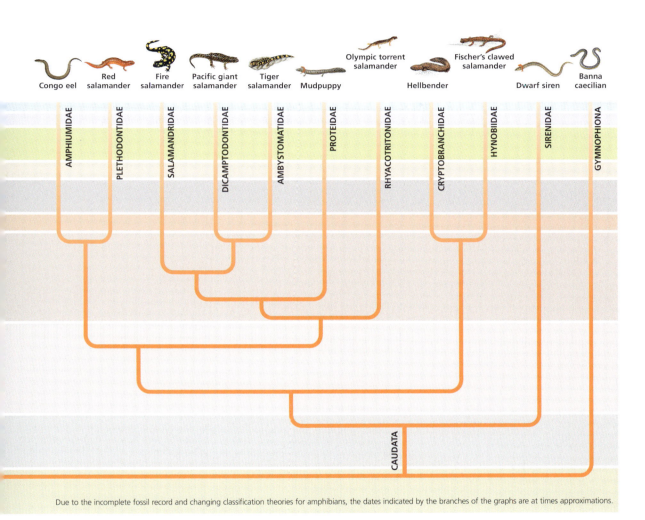

Poison frog in bromeliad

Mexican axolotyl

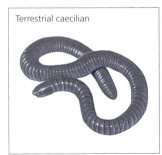

Terrestrial caecilian

Due to the incomplete fossil record and changing classification theories for amphibians, the dates indicated by the branches of the graphs are at times approximations.

Salamanders and Newts

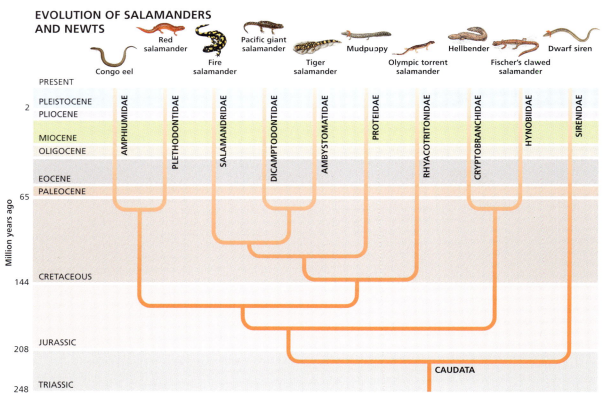

EVOLUTION OF SALAMANDERS AND NEWTS

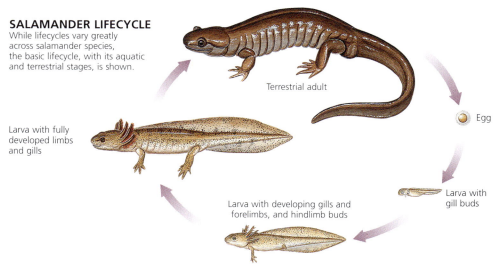

SALAMANDER LIFECYCLE
While lifecycles vary greatly across salamander species, the basic lifecycle, with its aquatic and terrestrial stages, is shown.

Terrestrial adult

Egg

Larva with gill buds

Larva with developing gills and forelimbs, and hindlimb buds

Larva with fully developed limbs and gills

FIRE SALAMANDER

FIRE SALAMANDER SKELETON

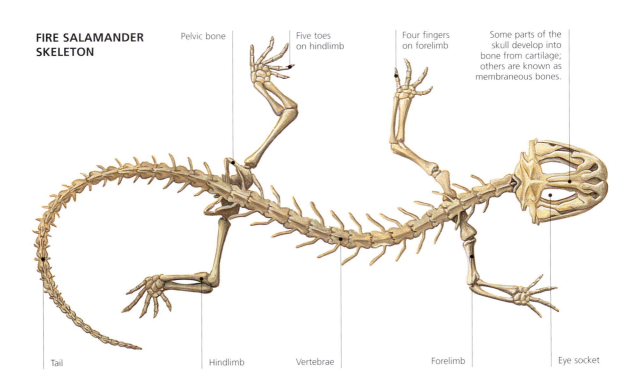

Salamanders and Newts

FIRE SALAMANDER
Salamandra salamandra

FISCHER'S CLAWED SALAMANDER
Onychodactylus fischeri

BLUE-SPOTTED SALAMANDER
Ambystoma laterale

PACIFIC GIANT SALAMANDER
Dicamptodon ensatus

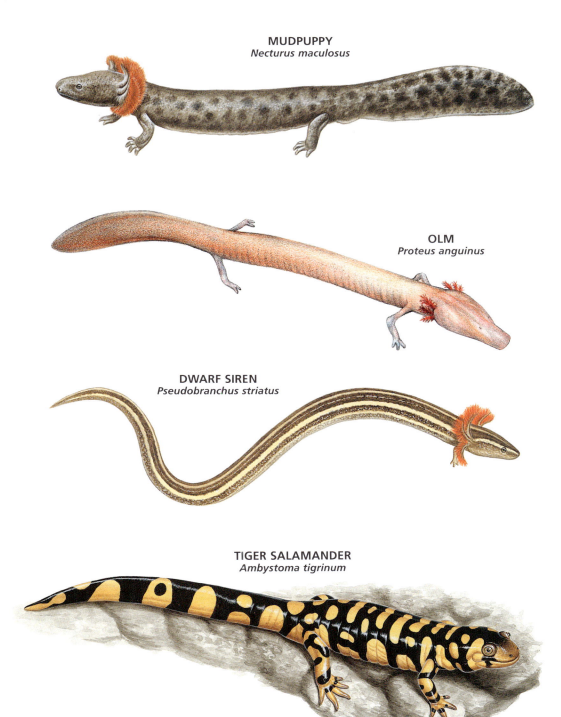

Salamanders and Newts

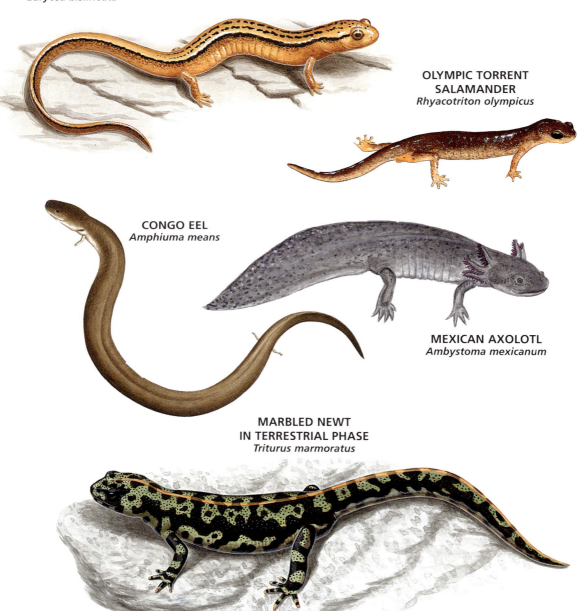

TWO-LINED SALAMANDER
Eurycea bislineata

OLYMPIC TORRENT SALAMANDER
Rhyacotriton olympicus

CONGO EEL
Amphiuma means

MEXICAN AXOLOTL
Ambystoma mexicanum

MARBLED NEWT IN TERRESTRIAL PHASE
Triturus marmoratus

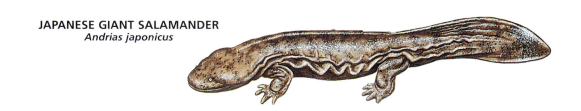

JAPANESE GIANT SALAMANDER
Andrias japonicus

**ALPINE NEWT
IN AQUATIC PHASE**
Triturus alpestris

RED SALAMANDER
Pseudotriton ruber

HELLBENDER
Cryptobranchus alleganiensis

Caecilians

Location
■ Caecilians

RINGED CAECILIAN
Siphonops annulatus

BANNA CAECILIAN
Ichthyophis bannanicus

Underdeveloped, weak eyes

Caecilians are the only legless amphibians.

Most caecilians live in underground tunnels.

Annuli (rings) partly encircle the body.

ADAPTATIONS FOR BURROWING

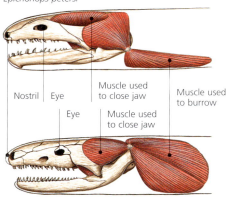

Family *Rhinatrematidae*
Epicrionops petersi

Nostril · Eye · Muscle used to close jaw · Muscle used to burrow

Eye · Muscle used to close jaw

Family *Ichthyophiidae*
Ichthyophis glutinosus

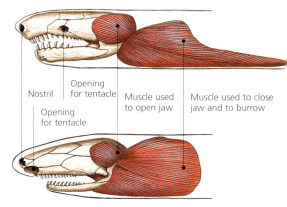

Family *Caecilidae*
Microcaecilia rabei

Nostril · Opening for tentacle · Muscle used to open jaw · Muscle used to close jaw and to burrow

Opening for tentacle

Family *Scolecomorphidae*
Crotaphatrema lamottei

AQUATIC CAECILIAN
Typhlonectes natans

SOUTHEAST ASIAN CAECILIAN
Ichthyophis kohtaoensis

TERRESTRIAL CAECILIAN
Dermophis mexicanus

SAO TOMÉ CAECILIAN
Schistometopum thomense

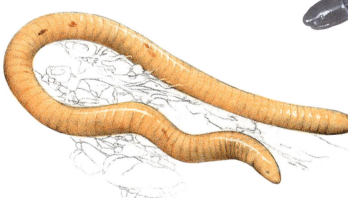

CAYENNE CAECILIAN
Typhlonectes compressicauda

Frogs and Toads

EVOLUTION OF FROGS AND TOADS

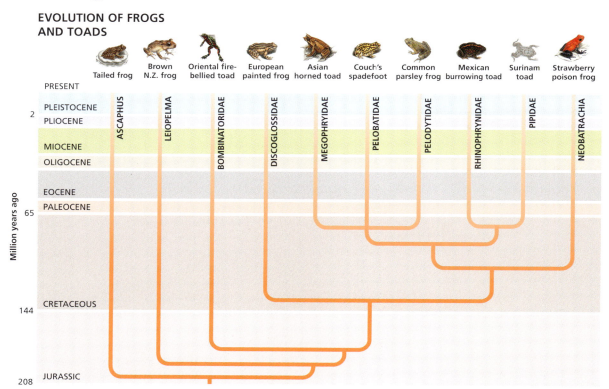

TREEFROG

TREEFROG SKELETON

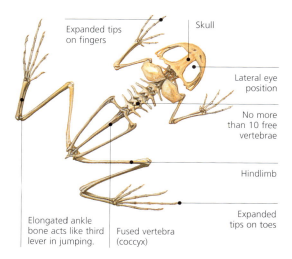

Primitive Frogs

Location
- Primitive frogs
- Pipids and transitional frogs
- Primitive frogs, pipids and transitional frogs

YELLOW-BELLIED TOAD
Bombina variegata

SURINAM TOAD
Pipa pipa

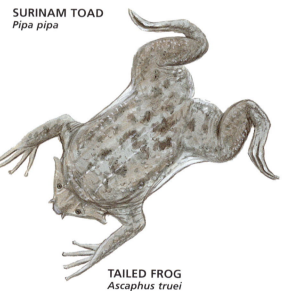

ASIAN HORNED TOAD
Megophrys nasuta

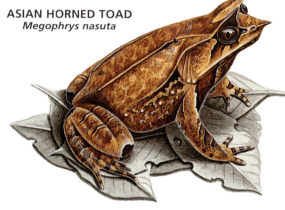

TAILED FROG
Ascaphus truei

MEXICAN BURROWING TOAD
Rhinophrynus dorsalis

American and Australasian Frogs

Location

- Leptodactylid frogs (American)
- Myobatrachid frogs (Australasian)

SOUTHERN PLATYPUS FROG
Rheobatrachus silus

GIANT BULLFROG
Limnodynastes interioris

CRUCIFIX TOAD
Notaden bennettii

WESTERN BARKING FROG
Eleutherodactylus augusti

WESTERN MARSH FROG
Heleioporus barycragus

CORROBOREE FROG
Pseudophryne corroboree

ORNATE HORNED TOAD
Ceratophrys ornata

SPOTTED GRASS FROG
Limnodynastes tasmaniensis

SOUTH AMERICAN BULLFROG
Leptodactylus pentadactylus

SCHMIDT'S FOREST FROG
Hydrolaetare schmidti

Toads, Treefrogs and Glass Frogs

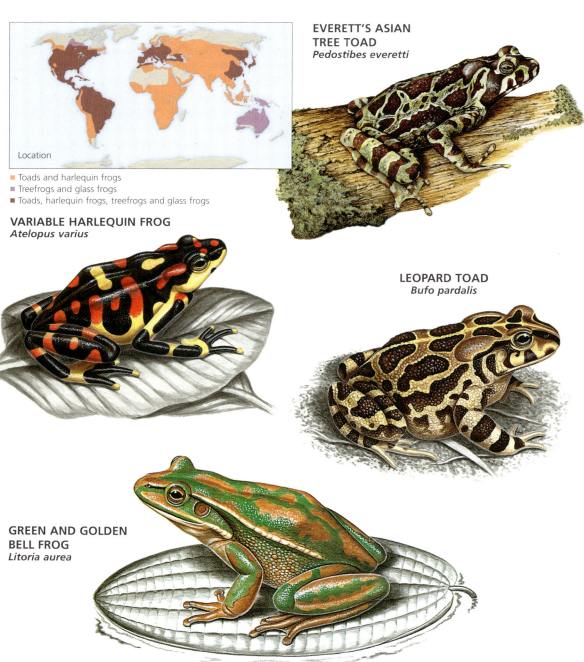

Location
- Toads and harlequin frogs
- Treefrogs and glass frogs
- Toads, harlequin frogs, treefrogs and glass frogs

EVERETT'S ASIAN TREE TOAD
Pedostibes everetti

VARIABLE HARLEQUIN FROG
Atelopus varius

LEOPARD TOAD
Bufo pardalis

GREEN AND GOLDEN BELL FROG
Litoria aurea

ASIATIC CLIMBING TOAD
Pedostibes hosii

DAINTY GREEN TREEFROG
Litoria gracilenta

GLASS FROG
Hylinobatrachium fleischmanni

BURROWING TREEFROG
Pternohyla fodiens

EUROPEAN GREEN TOAD
Bufo viridis

RED-EYED TREEFROG
Agalychnis callidryas

True Frogs and Poison Frogs

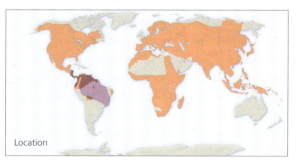

- True frogs
- Poison frogs
- True frogs and poison frogs

ORNATE BURROWING FROG
Hildebrandtia ornata

FUNEREAL POISON FROG
Phyllobates lugubris

SOLOMON ISLANDS TREEFROG
Platymantis guppyi

NORTHERN LEOPARD FROG
Rana sp.

ORANGE AND BLACK POISON FROG
Dendrobates leucomelas

Other Frogs and Toads

Location
■ Other frogs and toads

GOLD FROG
Brachycephalus ephippium

CAPE GHOST FROG
Heleophryne purcelli

SEYCHELLES FROG
Sooglossus sechellensis

ASIAN PAINTED FROG
Kaloula pulchra

SENEGAL RUNNING FROG
Kassina senegalensis

DARWIN'S FROG
Rhinoderma darwinii

WALLACE'S FLYING FROG
Rhacophorus nigropalmatus

EASTERN NARROW-MOUTHED TOAD
Gastrophryne carolinensis

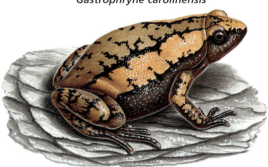

PARADOX FROG
Pseudis paradoxa

PAINTED REED FROG
Hyperolius marmoratus

RED-BANDED CREVICE CREEPER
Phrynomerus bifasciatus

FISHES

INTRODUCING FISHES

Classifying Fishes	426
Fish Characteristics	428
Evolution of Fishes	430

JAWLESS FISHES

Jawless Fishes	432

SHARKS

Sharks	434
Angel Sharks, Sawfishes and Dogfish Sharks	436
Ground Sharks	438
Mackerel Sharks	440
Carpetsharks, Bullhead Sharks, Frilled Sharks and Cow Sharks	442

RAYS AND CHIMAERAS

Rays and Chimaeras	444

BONY FISHES

Bony Fishes	448
Lungfishes and Coelacanth	450
Bichirs	452
Bonytongues	454
Eels	456
Sardines	458
Catfishes, Carps and Characins	460
Salmons	464
Dragonfishes and Lanternfishes	468
Cods and Anglerfishes	470
Spiny-rayed Fishes	472
Clingfishes, Flyingfishes, Killifishes, Ricefishes and Silversides	474
Oarfishes, Squirrelfishes and Dories	476
Pipefishes, Swampeels and Scorpionfishes	478
Perches, Groupers and Seabasses	480
Cichlids, Damselfishes, Wrasses, Parrotfishes and Blennies	482
Gobies, Flatfishes and Triggerfishes	484

Classifying Fishes

EVOLUTION OF FISHES

Introducing Fishes / FISHES

Million years ago

Period	Age (Mya)
PRESENT / PLEISTOCENE / PLIOCENE	2 / 5
MIOCENE	24
OLIGOCENE	34
EOCENE	54
PALEOCENE	65
CRETACEOUS	144
JURASSIC	208
TRIASSIC	248
PERMIAN	286
CARBONIFEROUS	360
DEVONIAN	408
SILURIAN	435
ORDOVICIAN	505

Groups shown: HYPEROARTIA, ACANTHOPTERYGII, PARACANTHOPTERYGII, SCOPELOMORPHA, CYCLOSQUAMATA, STENOPTERYGII, ESOCIFORMES, SALMONIFORMES, SILURIFORMES, GYMNOTIFORMES

Representative species: European brook lamprey, Leafy seadragon, Atlantic cod, Opah, Variegated lizardfish, Black dragonfish, Northern pike, Atlantic salmon, Striped catfish, South American knifefish

Barramundi

Flyingfish

Seahorse

426

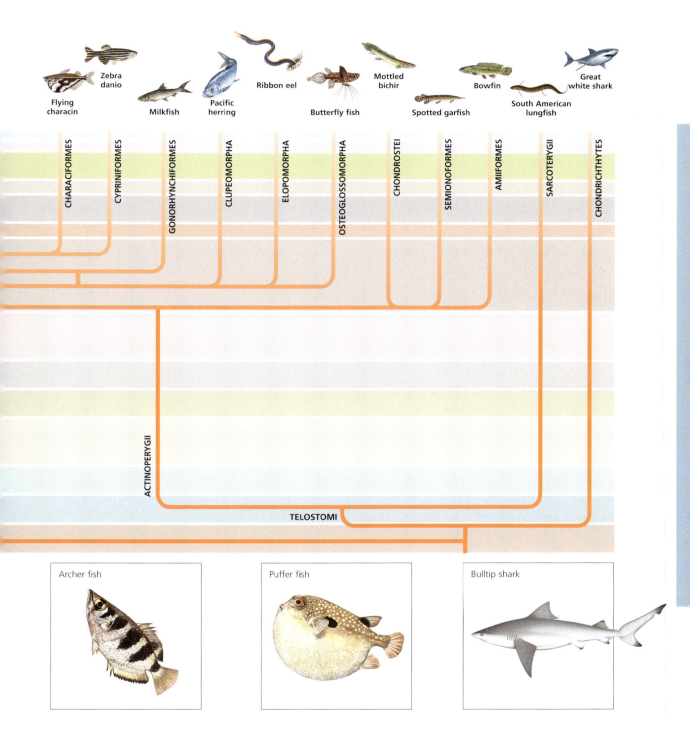

Fish Characteristics

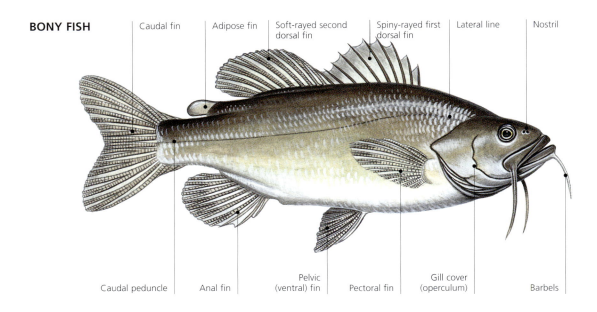

BONY FISH

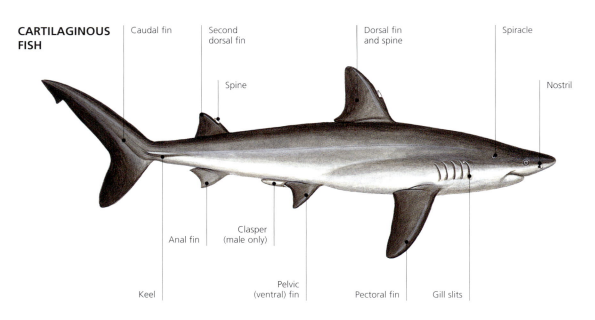

CARTILAGINOUS FISH

FISH SCALES

Ctenoid: bony fishes

Cycloid: bony fishes

Ganoid: armored fishes

Placoid: sharks

FISH RESPIRATION

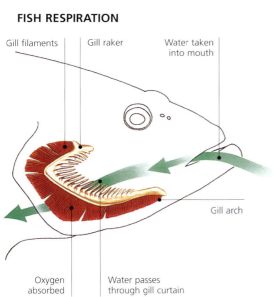

SHARK RESPIRATION

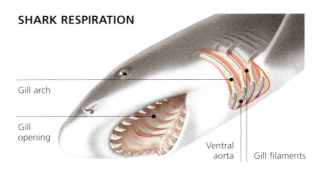
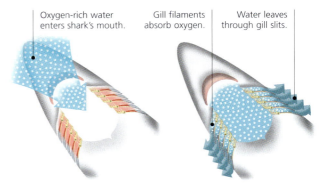

FISH TAILS

Butterflyfish

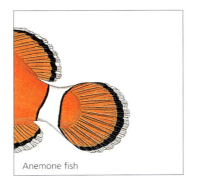
Anemone fish

Tuna

Evolution of Fishes

HEMICYCLASPIS
Ancestor of modern hagfishes,
it lived 500 million years ago.

CLIMATIUS
Earliest known group of jawed fishes,
it lived 450–400 million years ago.

DUNKLEOSTEUS
Early jawed fish, 8 feet (2 m) long,
it lived 370–360 million years ago.

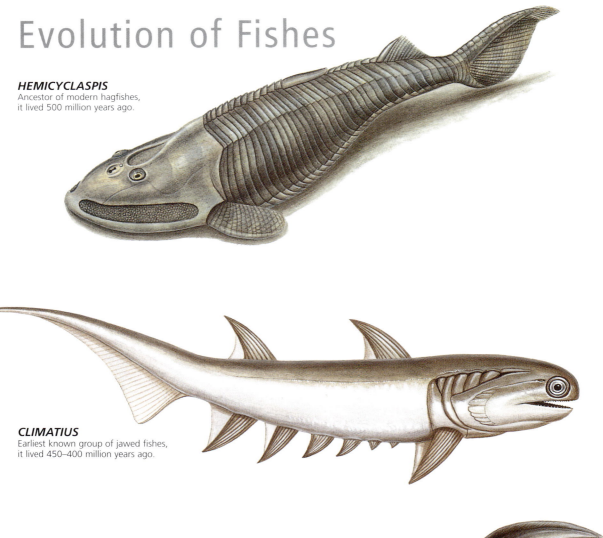

CLADOSELACHE
Ancestor of modern sharks, it lived 380 million years ago.

CHEIROLEPIS
Possibly an ancestor of sturgeons and bichirs, it lived 380 million years ago.

EUSTHENOPTERON
Lobe-finned fish with simple lungs, it lived 360 million years ago.

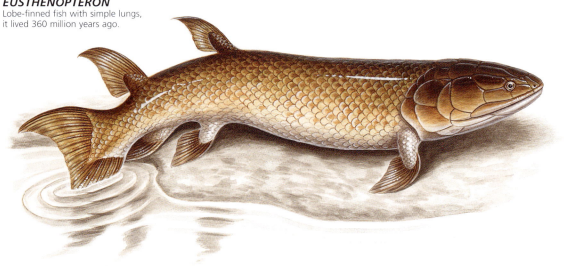

Jawless Fishes

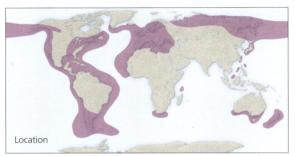

Location
■ Lampreys and hagfishes

SEA LAMPREY
Petromyzon marinus

EUROPEAN BROOK LAMPREY
Lampetra planeri

EUROPEAN RIVER LAMPREY
Lampetra fluviatilis

ATLANTIC HAGFISH
Myxine glutinosa

PACIFIC HAGFISH
Eptatretus stouti

HAGFISH KNOTS
By tying themselves in a knot, and bracing themselves against a carcass, limbless and jawless hagfishes gain enough leverage to tear off pieces of flesh.

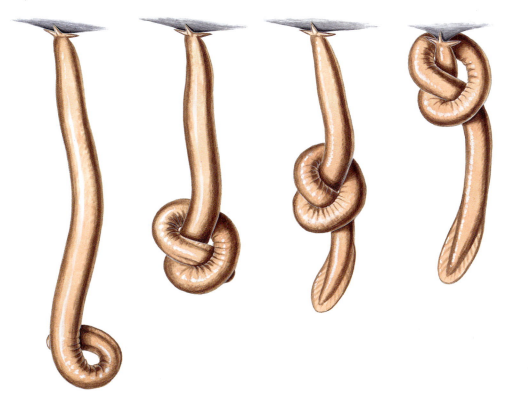

Sharks

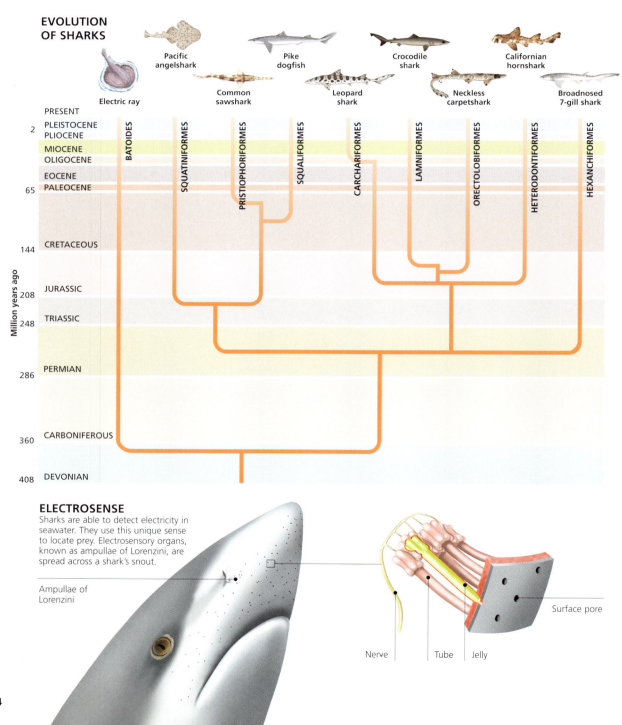

EVOLUTION OF SHARKS

ELECTROSENSE
Sharks are able to detect electricity in seawater. They use this unique sense to locate prey. Electrosensory organs, known as ampullae of Lorenzini, are spread across a shark's snout.

SHARK REPRODUCTION

Hornsharks
Hornsharks lay oval egg cases ringed with screwlike ridges. The mother wedges each case firmly into a crevice.

Swellsharks
Swellsharks lay flat, rectangular egg cases.

A growing shark and the yolk that feeds it.

The newborn shark begins to emerge.

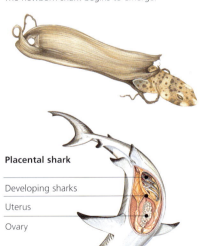

Placental shark
Developing sharks
Uterus
Ovary

TYPES OF TEETH

Blue shark

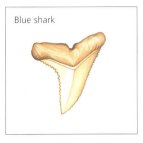

Tiger shark

Hornshark

Great white shark

Shortfin mako

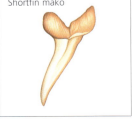

ANATOMY OF FEEDING

In its resting position, the shark's jaw lies just under its brain case.

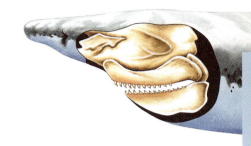

As the shark goes to grab its prey, the upper jaw slides forward. The lower jaw drops.

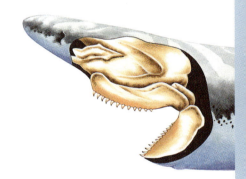

Muscle contractions force the jaw out of the mouth to give the shark a good grip.

Angel Sharks, Sawfishes and Dogfish Sharks

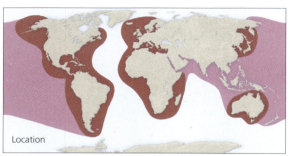
Location
- Sawfishes and dogfish sharks
- Sawfishes, dogfish sharks and angel sharks

SPINED PYGMY SHARK
Squaliolus laticaudus

COOKIE-CUTTER SHARK
Isistius brasiliensis

SMALLFIN GULPER SHARK
Centrophorus moluccensis

SPINY DOGFISH
Squalus acanthias

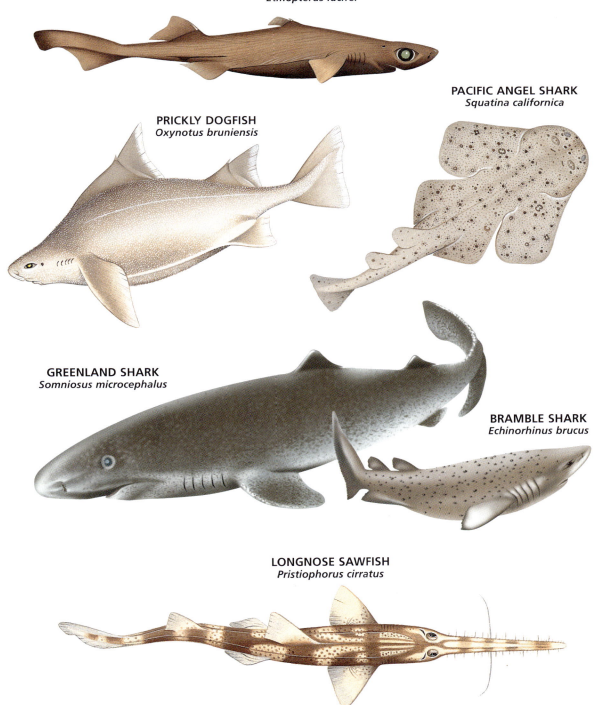

Ground Sharks

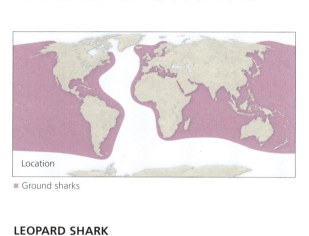
Location
■ Ground sharks

TYPES OF TAILS

Gray reef shark

Oceanic whitetip shark

Tiger shark

Atlantic weasel shark

Swell shark

LEOPARD SHARK
Triakis semifasciata

BLACKTIP REEF SHARK
Carcharhinus melanopterus

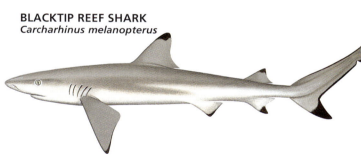

GRACEFUL CATSHARK
Proscyllium habereri

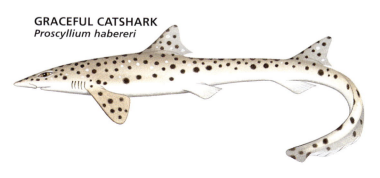

SCALLOPED HAMMERHEAD
Sphyrna lewini

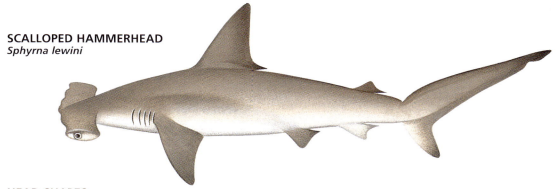

HEAD SHAPES

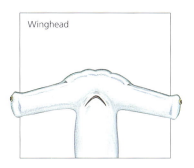

Winghead

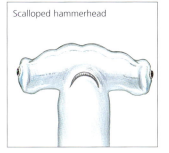

Scalloped hammerhead

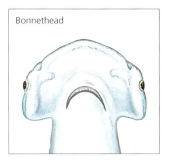

Bonnethead

HEAD HUNTING

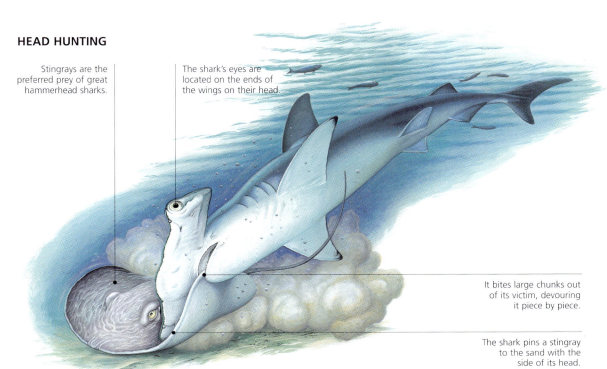

Stingrays are the preferred prey of great hammerhead sharks.

The shark's eyes are located on the ends of the wings on their head.

It bites large chunks out of its victim, devouring it piece by piece.

The shark pins a stingray to the sand with the side of its head.

Mackerel Sharks

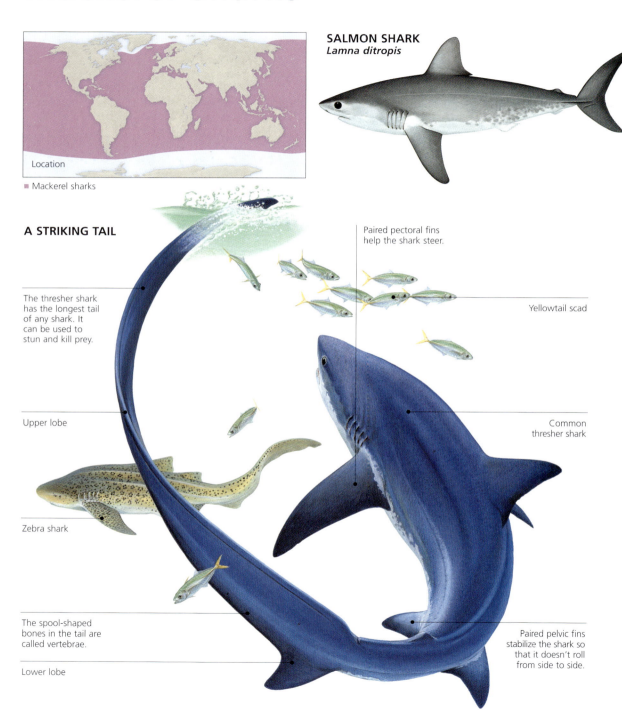

Location
■ Mackerel sharks

SALMON SHARK
Lamna ditropis

A STRIKING TAIL

The thresher shark has the longest tail of any shark. It can be used to stun and kill prey.

Upper lobe

Zebra shark

The spool-shaped bones in the tail are called vertebrae.

Lower lobe

Paired pectoral fins help the shark steer.

Yellowtail scad

Common thresher shark

Paired pelvic fins stabilize the shark so that it doesn't roll from side to side.

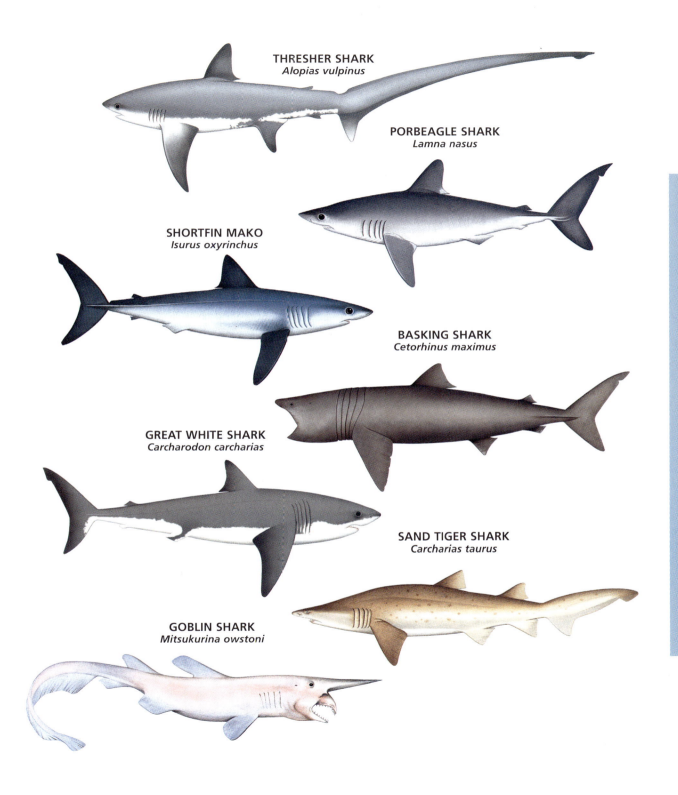

Carpetsharks, Bullhead Sharks, Frilled Sharks and Cow Sharks

Location
- Bullhead sharks, frilled sharks and cow sharks
- Bullhead sharks, frilled sharks, cow sharks and carpetsharks

WHALE SHARK FILTER FEEDING

Gill slits filter food from water.

Shark swallows filtered food.

Plankton, small fishes and water

CAMOUFLAGE
Some sharks have patterns on their skin that allow them to blend in with their environments. This helps conceal them from prey and predators.

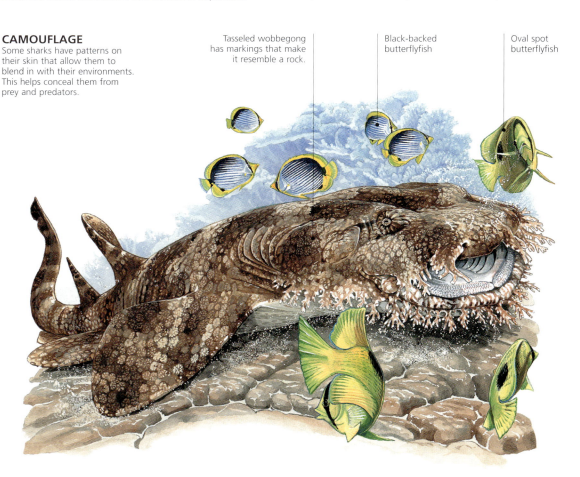

Tasseled wobbegong has markings that make it resemble a rock.

Black-backed butterflyfish

Oval spot butterflyfish

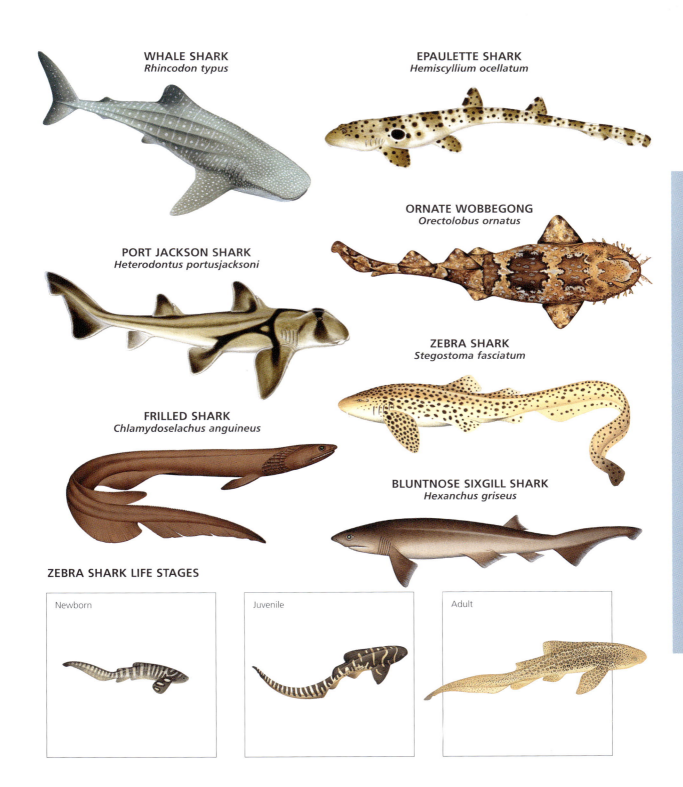

Rays and Chimaeras

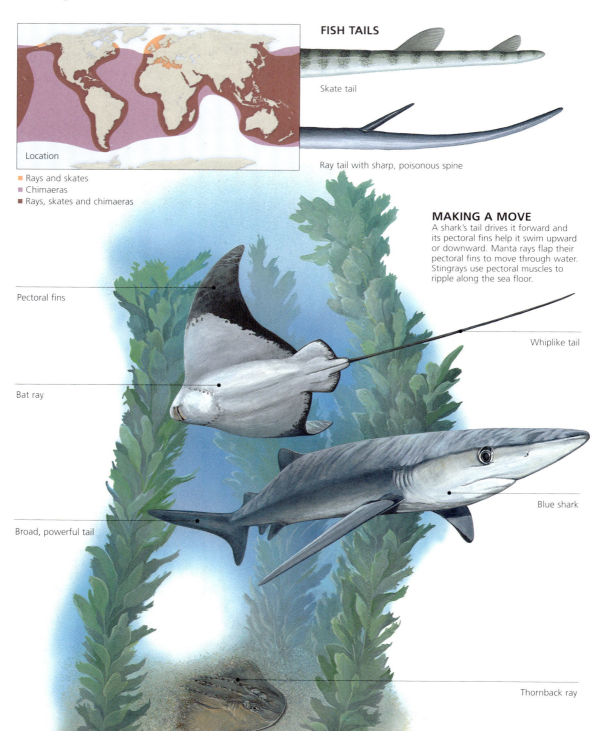

Location
- Rays and skates
- Chimaeras
- Rays, skates and chimaeras

FISH TAILS

Skate tail

Ray tail with sharp, poisonous spine

MAKING A MOVE
A shark's tail drives it forward and its pectoral fins help it swim upward or downward. Manta rays flap their pectoral fins to move through water. Stingrays use pectoral muscles to ripple along the sea floor.

Pectoral fins

Bat ray

Whiplike tail

Broad, powerful tail

Blue shark

Thornback ray

Spookfish

Atlantic devil ray

Freshwater sawfish snout

Shortnose chimaera

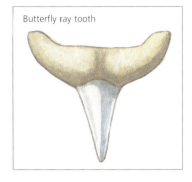
Butterfly ray tooth

Skate

Elephantfish

Stingray

Guitarfish

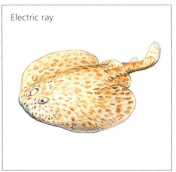
Electric ray

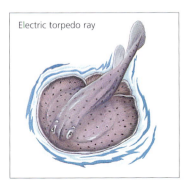
Electric torpedo ray

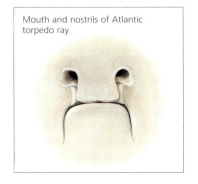
Mouth and nostrils of Atlantic torpedo ray

Rays and Chimaeras

BLUNT-NOSED CHIMAERA
Hydrolagus colliei

OCELLATED FRESHWATER STINGRAY
Potamotrygon motoro

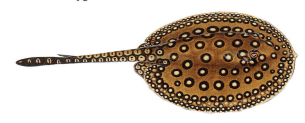

SIXGILL STINGRAY
Hexatrygon bickelli

BLIND ELECTRIC RAY
Typhlonarke aysoni

BLUE SKATE
Notoraja sp.

PORT DAVEY SKATE
Raja sp.

HOW THEY SWIM
Birds can overcome gravity because they don't weigh much for their size. The same is true of cartilaginous fishes. Cartilage is lighter than bone, and this allows sharks and rays to slice through the water with minimum effort.

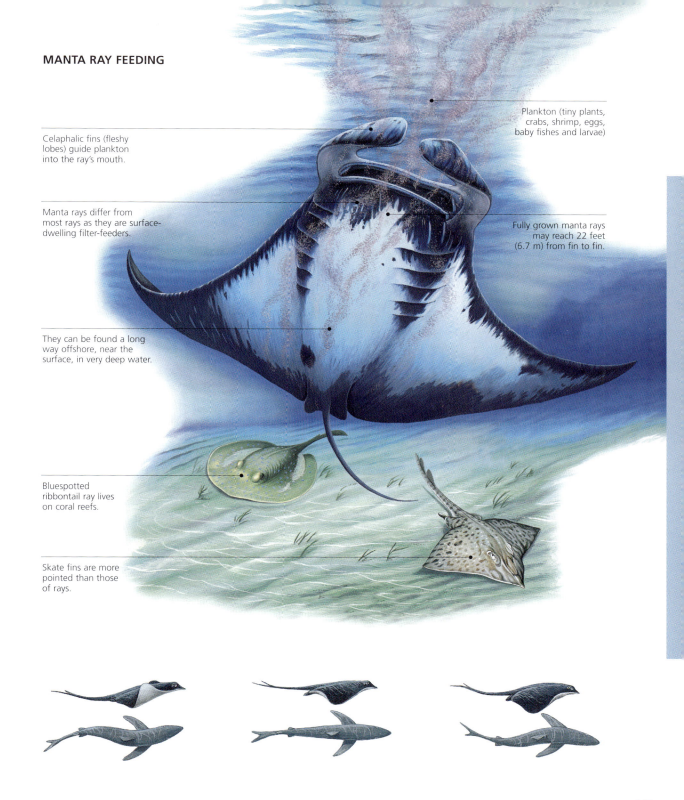

MANTA RAY FEEDING

Celaphalic fins (fleshy lobes) guide plankton into the ray's mouth.

Manta rays differ from most rays as they are surface-dwelling filter-feeders.

They can be found a long way offshore, near the surface, in very deep water.

Bluespotted ribbontail ray lives on coral reefs.

Skate fins are more pointed than those of rays.

Plankton (tiny plants, crabs, shrimp, eggs, baby fishes and larvae)

Fully grown manta rays may reach 22 feet (6.7 m) from fin to fin.

447

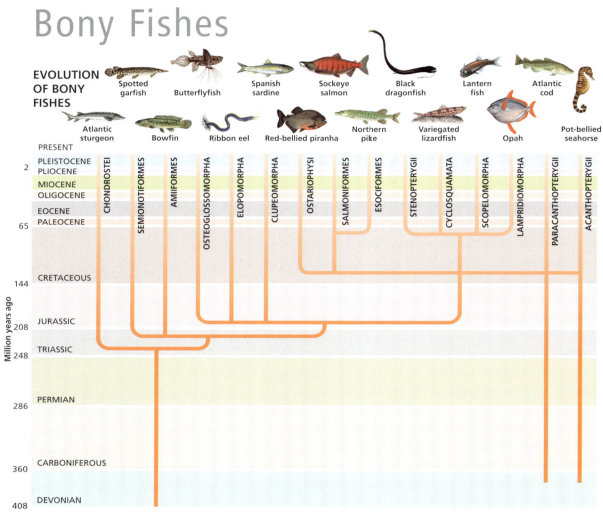

Bony Fishes

EVOLUTION OF BONY FISHES

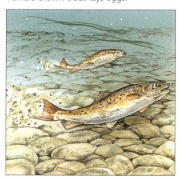
Female brown trout lays eggs.

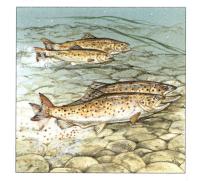
Eggs are fertilized by the male.

Hatchlings feed on orange yolk sac.

TROUT REPRODUCTION

BLUE-FIN TUNA

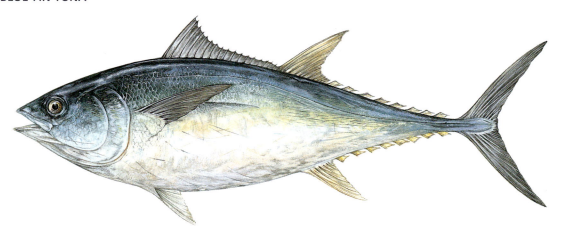

TUNA SKELETON

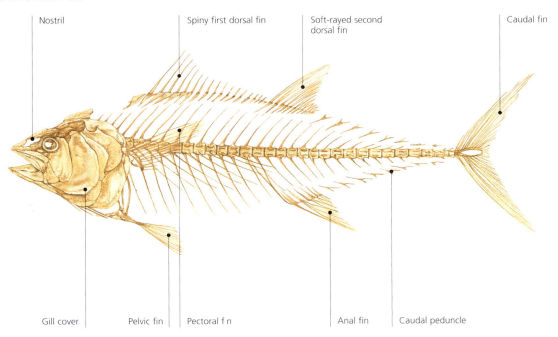

Lungfishes and Coelacanth

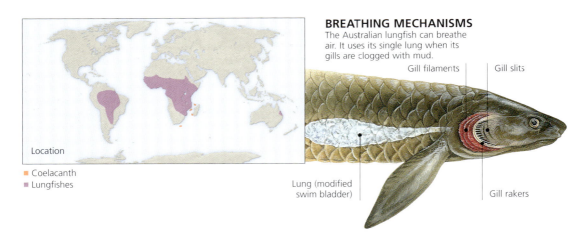

BREATHING MECHANISMS
The Australian lungfish can breathe air. It uses its single lung when its gills are clogged with mud.

Gill filaments · Gill slits · Gill rakers · Lung (modified swim bladder)

Location
- Coelacanth
- Lungfishes

SOUTH AMERICAN LUNGFISH
Lepidosiren paradoxa

AUSTRALIAN LUNGFISH
Neoceratodus forsteri

AFRICAN LUNGFISH
Protopterus dolloi

COELACANTH
Latimeria chalumnae

CRAWLING IN WATER
The African lungfish can swim and crawl. When crawling, it raises its body and propels itself forward with its thin, paired fins.

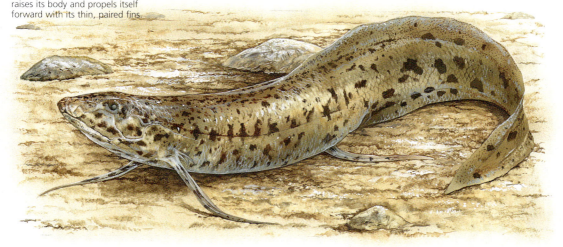

Bichirs

Location
■ Bichirs and relatives

MOTTLED BICHIR
Polypterus weeksi

ATLANTIC STURGEON
Acipenser oxyrinchus

BOWFIN
Amia calva

SPOTTED GAR
Lepisosteus oculatus

Bonytongues

Location

■ Bonytongues and relatives

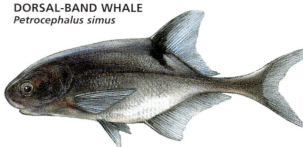

DORSAL-BAND WHALE
Petrocephalus simus

BUTTERFLYFISH
Pantodon buchholzi

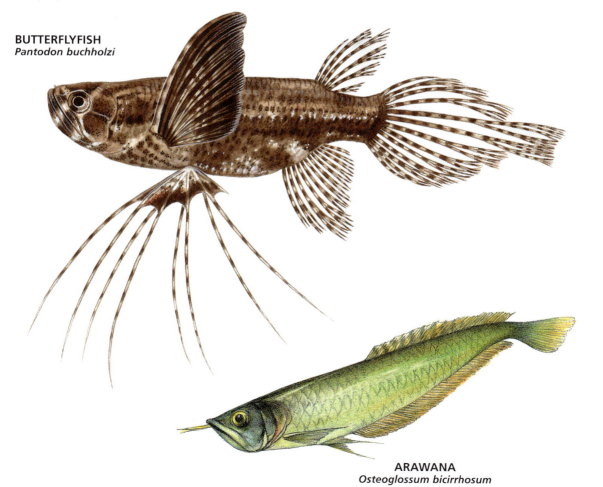

ARAWANA
Osteoglossum bicirrhosum

Eels

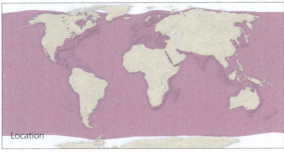
Location
■ Eels and relatives

Silver or mature

AMERICAN EEL
Anguilla rostrata

Yellow or juvenile

JUVENILE GARDEN EELS
Being able to burrow tail-first into the shell bottom of tropical seas, garden eels are adapted to life firmly anchored into the sea floor. They emerge from their burrows to feed on plankton.

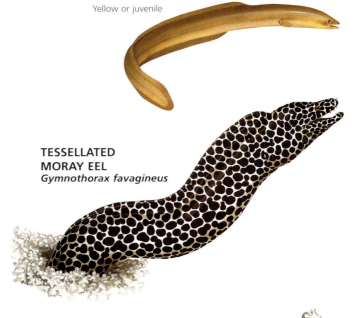

TESSELLATED MORAY EEL
Gymnothorax favagineus

RIBBON EEL
Rhinomuraena quaesita

UMBRELLA MOUTH GULPER EEL
Eurypharynx pelecanoides

Jaws nearly one-fourth the total body length

Tiny eye

Luminous organ emits a continuous pink glow punctuated by red flashes.

Tapering, whiplike tail

Hatchet fish

DEEPSEA DWELLERS
Gulper eels live in a dark, slow-moving, sparsely inhabited world over 6,500 feet (1,980 m) below the surface of the sea. Their hunt for prey is aided by a huge mouth, prickly teeth, and a luminous organ at the end of a long tail.

Sardines

Location
■ Sardines

ATLANTIC THREAD HERRING
Opisthonema oglinum

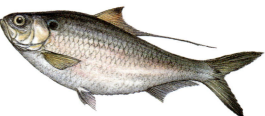

PERUVIAN ANCHOVETA
Engraulis ringens

AMERICAN GIZZARD SHAD
Dorosoma cepedianum

CASPIAN SPRAT
Clupeonella cultriventris

EUROPEAN PILCHARD
Sardina pilchardus

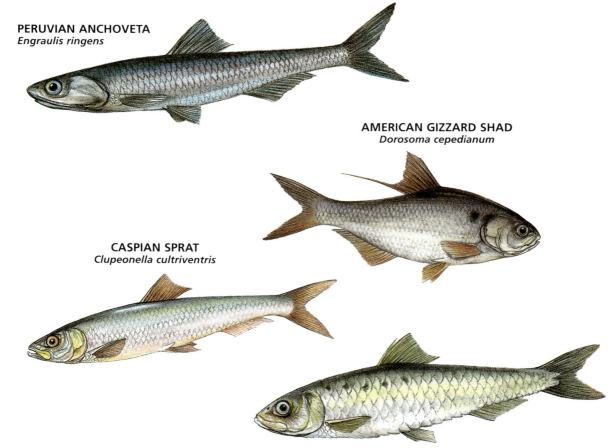

AMERICAN SHAD
Alosa sapidissima

BLACKFIN WOLF-HERRING
Chirocentrus dorab

SPANISH SARDINE
Sardinella aurita

TYPES OF SARDINES

Anchovies

Pacific herring

Sardines

Catfishes, Carps and Characins

Location
- Characins
- Catfishes and carps
- Characins, catfishes and carps

MILKFISH
Chanos chanos

BITTERLING
Rhodeus sericeus

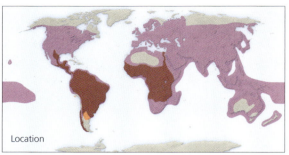

Breeding habits
Female bitterling lays eggs inside a mussel. The male sheds sperm over the mussel's syphon. As the mussel breathes, the sperm is drawn into the cavity to fertilize the eggs.

SLIMY LOACH
Acantophthalmus myersi

EGG INCUBATION
The spotted African squeaker begins life in the mouth of another species of fish. A female cichlid picks up the eggs and broods them in her mouth.

A LEAP FOR LIFE
Splash tetra leap out of the water and spawn on the underside of objects such as leaves. The male splashes water on the eggs with his tail until they hatch.

ELECTRIC CATFISH
Malapterurus electricus

SADDLED HILLSTREAM LOACH
Homaloptera orthogoniata

TRICOLOR SHARKMINNOW
Balantiocheilos melanopterus

Catfishes, Carps and Characins

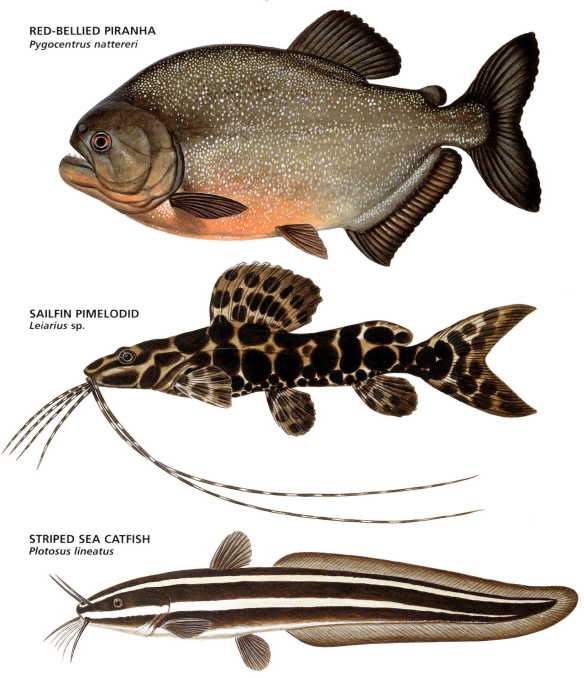

RED-BELLIED PIRANHA
Pygocentrus nattereri

SAILFIN PIMELODID
Leiarius sp.

STRIPED SEA CATFISH
Plotosus lineatus

MARBLED HATCHETFISH
Carnegiella strigata

STRIPED HEADSTANDER
Anostomus anostomus

ZEBRA DANIO
Danio rerio

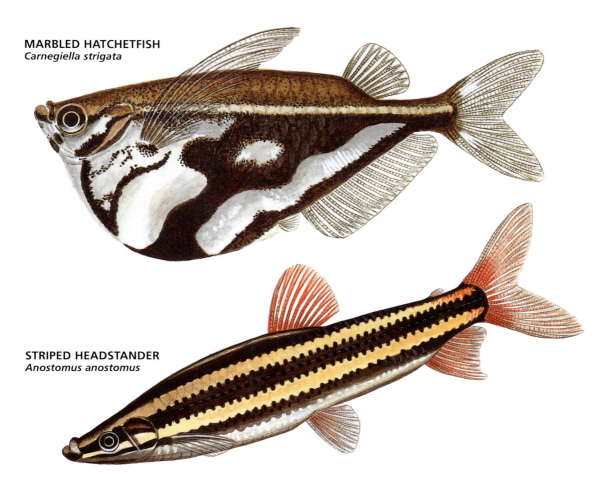

Salmons

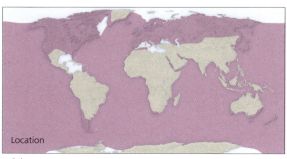
Location
■ Salmons

PIKE HEAD SKELETON
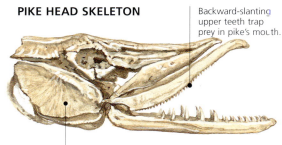
Backward-slanting upper teeth trap prey in pike's mouth.
Large head enables pike to swallow prey almost half its size.

SOCKEYE SALMON LIFE STAGES

Parr lives in rivers and small lakes.

It enters the sea as a smolt.

Smolt matures into a sea-going adult.

ARTHUR'S PARAGALAXIAS
Paragalaxias mesotes

CISCO
Coregonus artedi
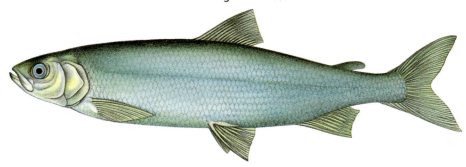

SOCKEYE SALMON LIFECYCLE

Summer–Autumn
Female buries several thousand fertilized eggs in a shallow nest dug in the gravel of a river bed.

Summer–Autumn
During spawning the male turns a brilliant red and develops a long, hooked jaw. After spawning, the salmon soon die.

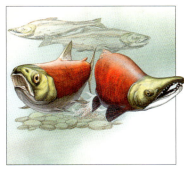

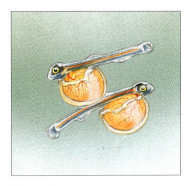

Winter–Spring
Eggs take several months to hatch. Hatchlings remain in the gravel until the attached yolk has been consumed.

Spring
Salmon fry wriggle out into the stream, inhabiting a lake for a year before beginning the journey out to sea.

■ Location

■ Migratory path

Spring
The sea-bound salmon, known as smolts, face many predators such as other fishes, birds and mammals.

Spring–Summer
Returning to their birth river, some are eaten by brown bears. The survivors begin to show dramatic spawning colors.

Spring–Summer
After years at sea, salmon approach sexual maturity and are ready to face the rigors of the spawning migration.

Salmons

RAINBOW TROUT
Oncorhynchus mykiss

GOLDEN TROUT
Oncorhynchus aguabonita

LAKE TROUT
Salvelinus namaycush

GRAYLING
Thymallus thymallus

NORTHERN PIKE
Esox lucius

ALASKA BLACKFISH
Dallia pectoralis

SOCKEYE SALMON
Oncorhynchus nerka

ATLANTIC SALMON
Salmo salar

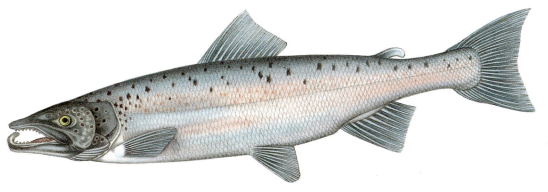

Dragonfishes and Lanternfishes

DRAGONFISH
Family *Stomiidae*

Location
■ Dragonfishes, lanternfishes and lizardfishes

BRISTLEMOUTH
Family *Gonostomatidae*

CROSS-TOOTHED PERCH
Order *Stomiiformes*

LANTERNFISH
Family *Myctophidae*

VARIEGATED LIZARDFISH
Synodus variegatus

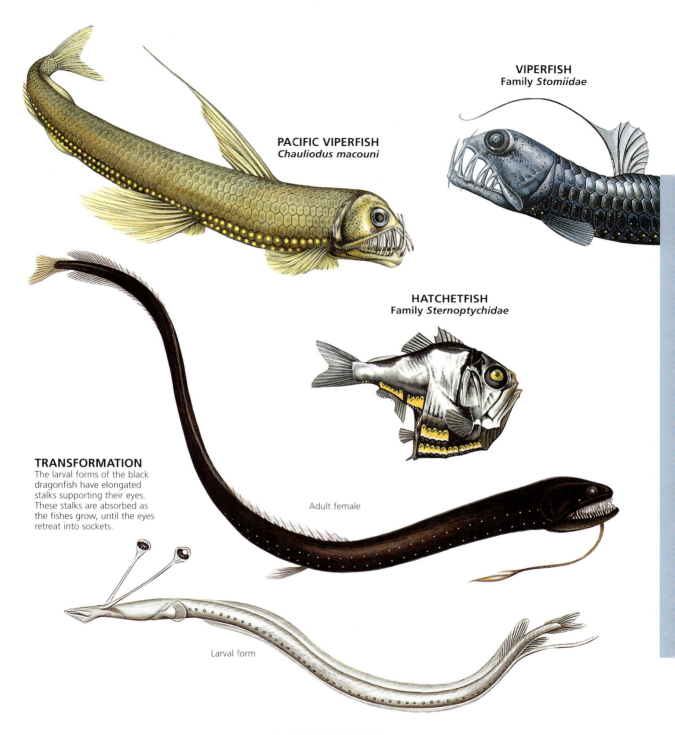

PACIFIC VIPERFISH
Chauliodus macouni

VIPERFISH
Family *Stomiidae*

HATCHETFISH
Family *Sternoptychidae*

TRANSFORMATION
The larval forms of the black dragonfish have elongated stalks supporting their eyes. These stalks are absorbed as the fishes grow, until the eyes retreat into sockets.

Adult female

Larval form

BLACK DRAGONFISH
Idiacanthus fasciola

Cods and Anglerfishes

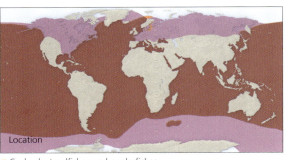

Location
- Cuskeels, toadfishes and anglerfishes
- Cods and troutperches
- Cuskeels, toadfishes, anglerfishes, cods and troutperches

ATLANTIC COD
Gadus morhua

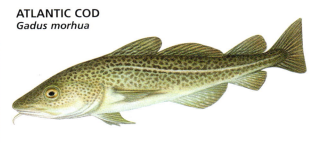

PIRATE PERCH
Aphredoderus sayanus

SPLENDID TOADFISH
Sanopus splendidus

PAXTON'S WHIPNOSE ANGLER
Gigantactis paxtoni

CIRCUMPOLAR BURBOT
Lota lota

PARASITIC LIFESTYLE
Male anglerfishes are tiny compared to females, with toothless jaws and no baits to lure prey. Males sometimes attach themselves to females and live as parasites.

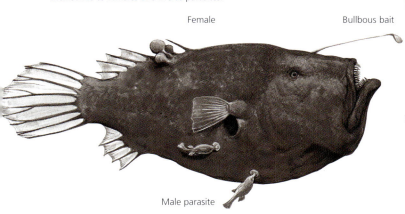

Female

Bullous bait

Male parasite

CAMOUFLAGE TECHNIQUES
Bottom-dwelling anglerfishes blend in with their environment to snare prey. Frogfishes can change color and hide among coral.

Roughjaw frogfish

Sargassum fish

Spiny-rayed Fishes

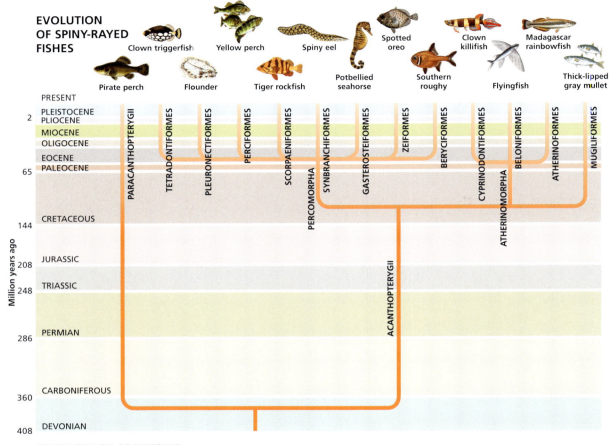

STICKLEBACK COURTSHIP

Male uses his bright colors to attract a passing female.

The pair begins a courtship dance.

Male entices female to a nest he has built.

AUSTRALIAN RAINBOWFISH

RAINBOWFISH SKELETON

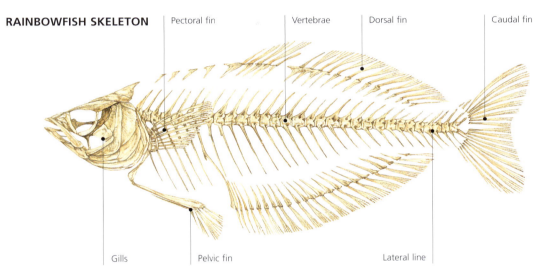

Pectoral fin | Vertebrae | Dorsal fin | Caudal fin
Gills | Pelvic fin | Lateral line

He guides and coaxes her inside.

Once inside, the female lays her eggs, then leaves.

Male fertilizes eggs and guards them until they hatch.

Clingfishes, Flyingfishes, Killifishes, Ricefishes and Silversides

Location
- Killifishes, ricefishes and silversides
- Clingfishes and flyingfishes
- Killifishes, ricefishes, silversides, clingfishes and flyingfishes

MANDARIN FISH
Synchiropus splendidus

SWORDTAIL
Xiphophorus helleri

Changing sexes
The male swordtail is smaller than the female, and has a swordlike extension to its tail. Females sometimes turn into males, even after giving birth.

Female

Male

URCHIN CLINGFISH
Diademichthys lineatus

FLYINGFISH
Family *Exocoetidae*

MALAYAN HALFBEAK
Dermogenys pusilla

Oarfishes, Squirrelfishes and Dories

Location
- Oarfishes and squirrelfishes
- Oarfishes, squirrelfishes and dories

FLASHING ON AND OFF
Flashlight fishes have light organs which contain luminous bacteria. To hide from predators, the fish covers the light organ with a type of eyelid called a melanphore.

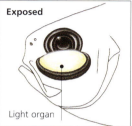
Exposed — Light organ

Hidden — Melanphore

FLASHLIGHT FISH
Photoblepharon palpebratus

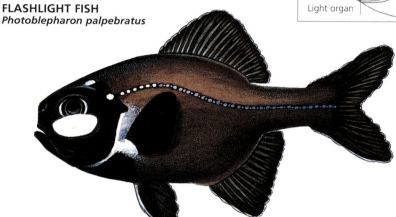

OARFISH
Regalecus glesne

RIBBONFISH
Trachipterus sp.

SOUTHERN ROUGHY
Trachichthys australis

OPAH
Lampris guttatus

AUSTRALIAN PINEAPPLEFISH
Cleidopus gloriamaris

SPOTTED OREO
Pseudocyttus maculatus

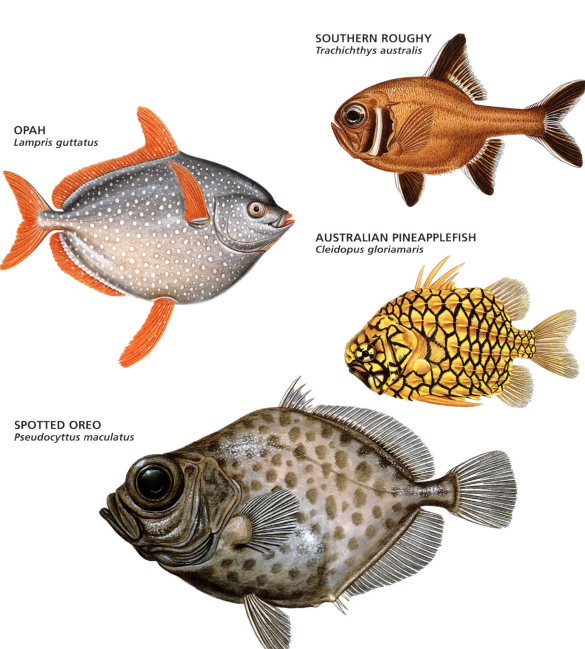

Pipefishes, Swampeels and Scorpionfishes

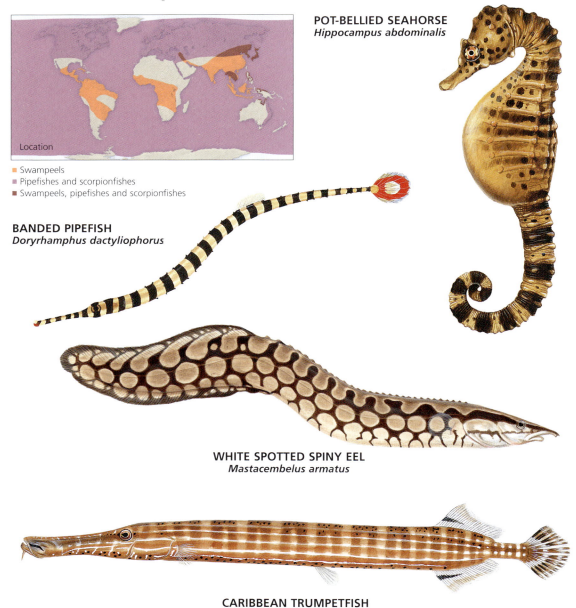

Location
- Swampeels
- Pipefishes and scorpionfishes
- Swampeels, pipefishes and scorpionfishes

POT-BELLIED SEAHORSE
Hippocampus abdominalis

BANDED PIPEFISH
Doryrhamphus dactyliophorus

WHITE SPOTTED SPINY EEL
Mastacembelus armatus

CARIBBEAN TRUMPETFISH
Aulostomus maculatus

SEAHORSE REPRODUCTION

The female lays eggs in the male's marsupial-like pouch, leaving them in his care. The eggs are incubated in his pouch until they hatch.

LUMPSUCKER
Cyclopterus lumpus

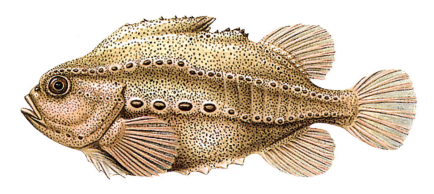

TIGER ROCKFISH
Sebastes nigrocinctus

COCKATOO WASPFISH
Ablabys taenionotus

Perches, Groupers and Seabasses

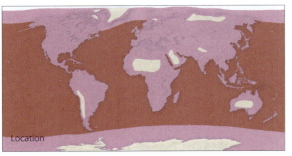

Location
- Perches
- Perches, groupers and seabasses

SPLENDID LICORICE GOURAMI
Parosphromenus dreissneri

SIXLINE SOAPFISH
Grammistes sexlineatus

PURPLEQUEEN
Pseudanthis tuka

COMMON DOLPHINFISH
Coryphaena hippurus

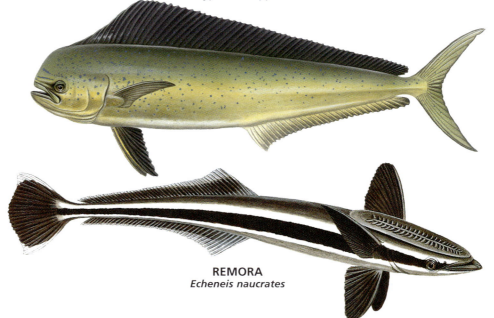

REMORA
Echeneis naucrates

GIANT GROUPER
Epinephelus lanceolatus

SOUTH-EAST ASIA PIKEHEAD
Luciocephalus pulcher

MEYER'S BUTTERFLYFISH
Chaetodon meyeri

MOORISH IDOL
Zanclus cornutus

Cichlids, Damselfishes, Wrasses, Parrotfishes and Blennies

Location
- Cichlids, damselfishes, wrasses and parrotfishes
- Cichlids, damselfishes, wrasses, parrotfishes and blennies

MUDSKIPPERS
The male protects eggs laid by the female by wrapping his body around them.

CHANGING COLOR AND SEX
As they mature, highfin parrotfishes (*Scarus altipinnis*) travel through three color phases, which also indicate changes to gender.

Juvenile: usually asexual female

Initial phase: usually female

Terminal phase: always a mature male

HARLEQUIN TUSKFISH
Choerodon fasciatus

STRIPED JULIE
Julidochromis regani

BLACK-HEADED BLENNY
Lipophrys nigriceps

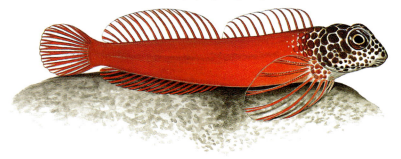

Gobies, Flatfishes and Triggerfishes

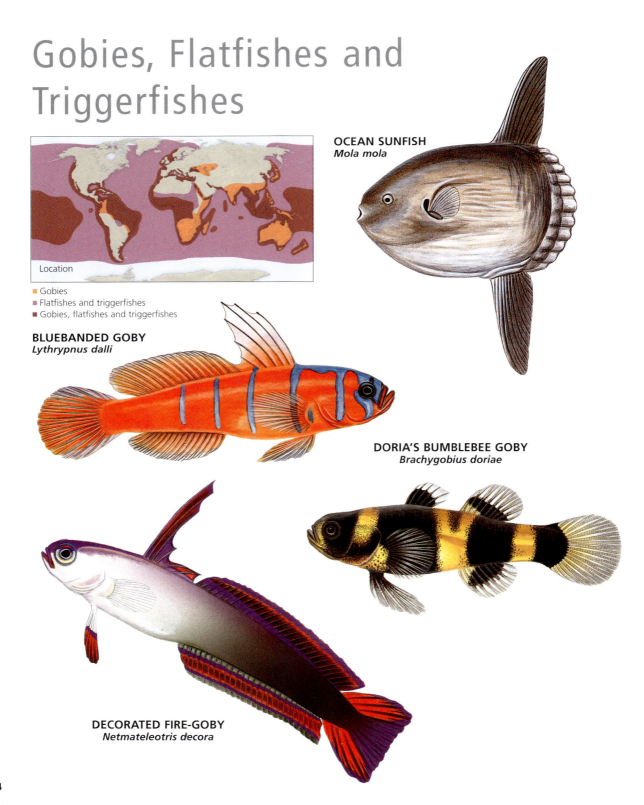

Location
- Gobies
- Flatfishes and triggerfishes
- Gobies, flatfishes and triggerfishes

OCEAN SUNFISH
Mola mola

BLUEBANDED GOBY
Lythrypnus dalli

DORIA'S BUMBLEBEE GOBY
Brachygobius doriae

DECORATED FIRE-GOBY
Netmateleotris decora

RIGHT-EYED FLATFISH
The left eye of right-eyed flatfishes (flounders) moves toward the right eye. The front of the skull twists to bring the jaws sideways.

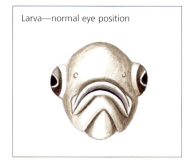
Larva—normal eye position

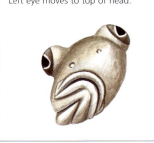
Left eye moves to top of head.

Adult—both eyes on right side

PEACOCK FLOUNDER
Bothus lunatus

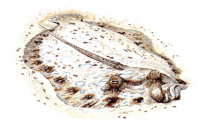

PUFFERFISH
Pufferfishes inflate themselves into a spiny globe when threatened.

Relaxed

Threatened

SURVIVAL TECHNIQUES
The mimic filefish (*Paraluteres prionurus*) has evolved to look like the toxic blacksaddled puffer (*Canthigaster valentini*).

Blacksaddled puffer (venomous)

Mimic filefish (non-venomous)

FIGURE-EIGHT PUFFER
Tetraodon biocellatus

CLOWN TRIGGERFISH
Balistoides conspicillum

INVERTEBRATES

INTRODUCING INVERTEBRATES

Classifying Invertebrates 490

INVERTEBRATE CHORDATES, SPONGES AND CNIDARIANS

Invertebrate Chordates,
   Sponges and Cnidarians 492

MOLLUSKS

Mollusks 494
Squids and Octopuses 496

WORMS

Worms 498

ECHINODERMS

Echinoderms 500

ARTHROPODS

Arthropods 502
Arachnids 504
Spiders 506
Orb-weaving Spiders 510
Scorpions, Mites and Ticks 512
Crustaceans, Silverfishes,
   Centipedes and Millipedes 514
Insects 518
Dragonflies, Mayflies
   and Mantids 520
Cockroaches, Termites
   and Lice 522
Crickets and Grasshoppers 524
Bugs, Lacewings and Thrips 526
Beetles 530
Ladybugs 534
Flies, Fleas and Mosquitoes 536
Butterflies and Moths 540
Butterflies 542
Moths 544
Bees, Wasps and Ants 546
Wasps and Ants 548
Honey Bees 550

Classifying Invertebrates

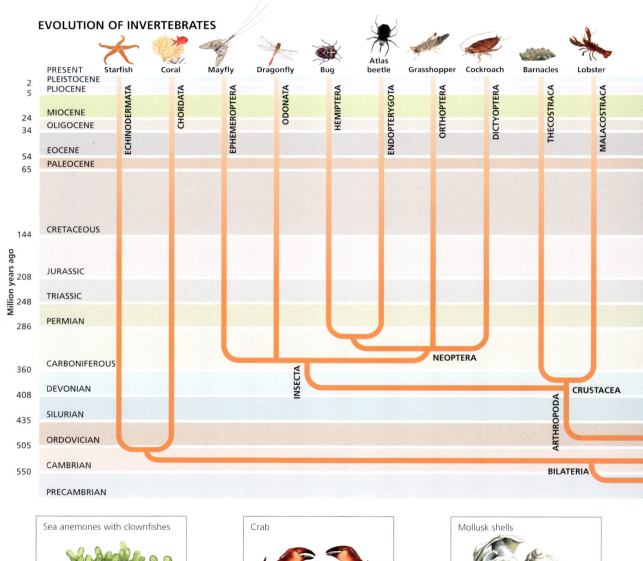

EVOLUTION OF INVERTEBRATES

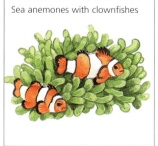

Sea anemones with clownfishes

Crab

Mollusk shells

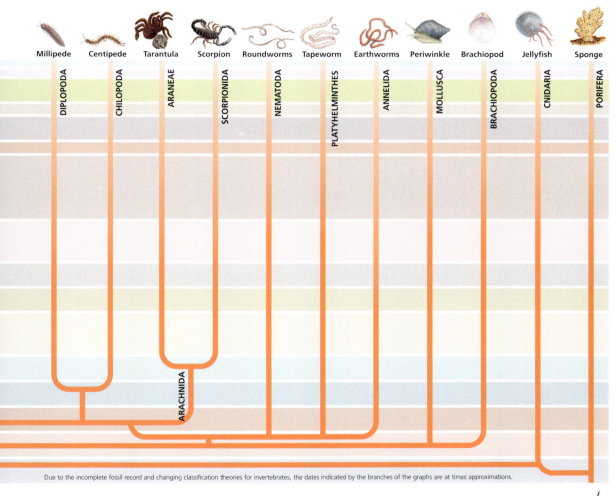

Weevil

Butterfly

Redback spider

Invertebrate Chordates, Sponges and Cnidarians

CORAL REEF ECOSYSTEM
Corals thrive in the shallow waters of the tropical Pacific and Indian oceans. Thousands of individual coral polyps, each protected by a limestone exoskeleton, group together to form living colonies; these in turn form coral reefs.

The crown-of-thorns starfish attacks and eats corals by disgorging its stomach over a colony, then absorbing the liquefied tissues.

Coral reefs provide shelter, food and breeding territory for thousands of species of plants and animals, such as fish, sharks and turtles.

Within certain coral communities, individual polyps have specialized functions—feeding, breeding or defense.

COLONIAL SEA SQUIRT
Didemnum molle

SPONGE
Spongia officinalis

SOFT CORAL
Lophelia pertusa

SEA SQUIRT
Class *Urochordata*

Sea anemone

Sea sponge

Portuguese man o'war

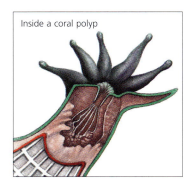

Inside a coral polyp

Jellyfish

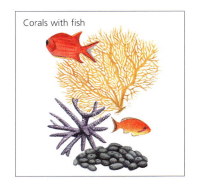

Corals with fish

Mollusks

MUSSEL LIFECYCLE

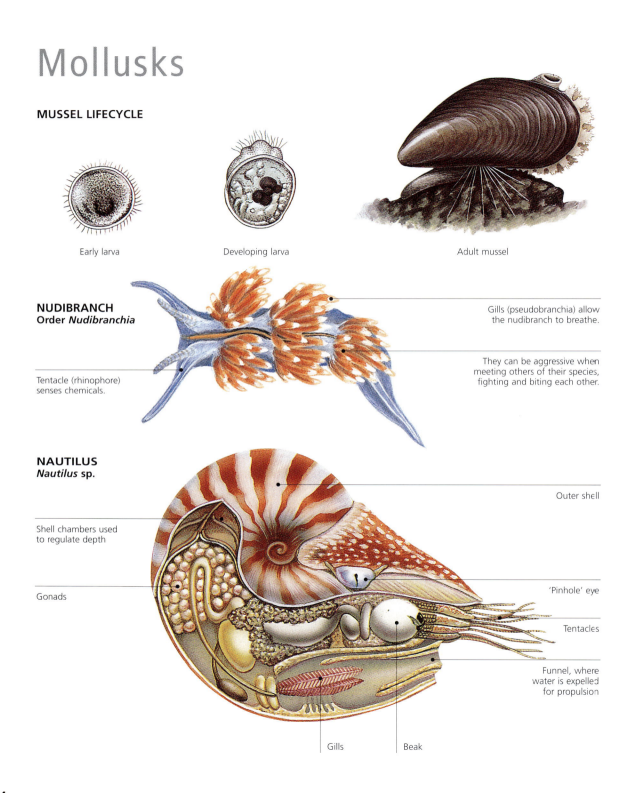

Early larva

Developing larva

Adult mussel

NUDIBRANCH
Order *Nudibranchia*

Tentacle (rhinophore) senses chemicals.

Gills (pseudobranchia) allow the nudibranch to breathe.

They can be aggressive when meeting others of their species, fighting and biting each other.

NAUTILUS
Nautilus sp.

Shell chambers used to regulate depth

Gonads

Outer shell

'Pinhole' eye

Tentacles

Funnel, where water is expelled for propulsion

Gills

Beak

Common garden snail

Periwinkle (marine snail)

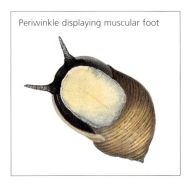
Periwinkle displaying muscular foot

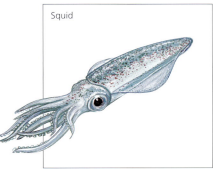
Squid

Oyster shells

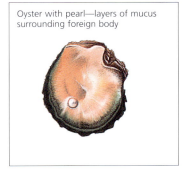
Oyster with pearl—layers of mucus surrounding foreign body

Brachiopod shells

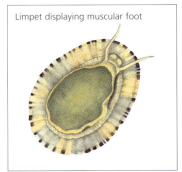
Limpet displaying muscular foot

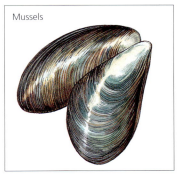
Mussels

Bivalve shells

Mud whelk

Common garden slug

Squids and Octopuses

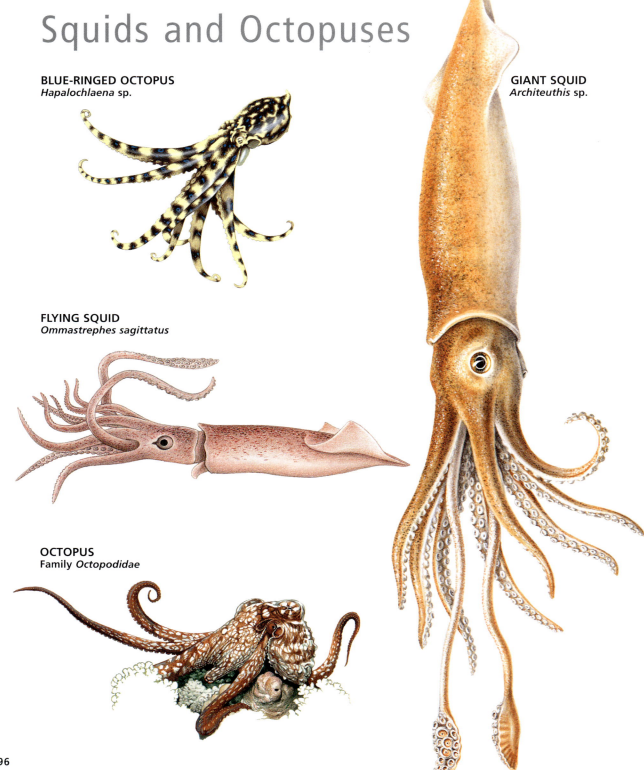

BLUE-RINGED OCTOPUS
Hapalochlaena sp.

GIANT SQUID
Architeuthis sp.

FLYING SQUID
Ommastrephes sagittatus

OCTOPUS
Family *Octopodidae*

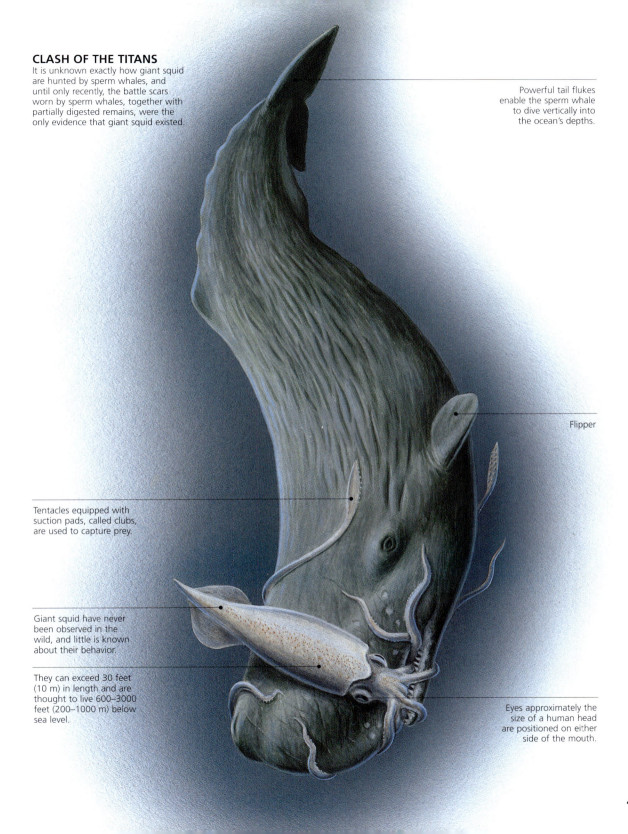

CLASH OF THE TITANS
It is unknown exactly how giant squid are hunted by sperm whales, and until only recently, the battle scars worn by sperm whales, together with partially digested remains, were the only evidence that giant squid existed.

Powerful tail flukes enable the sperm whale to dive vertically into the ocean's depths.

Flipper

Tentacles equipped with suction pads, called clubs, are used to capture prey.

Giant squid have never been observed in the wild, and little is known about their behavior.

They can exceed 30 feet (10 m) in length and are thought to live 600–3000 feet (200–1000 m) below sea level.

Eyes approximately the size of a human head are positioned on either side of the mouth.

Worms

PARASITIC WORMS

Hookworm eggs

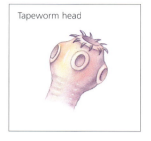
Tapeworm head

WHIPWORMS
Trichuris sp.

MARINE TAPEWORM
Class *Cestoidea*

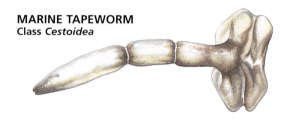

MARINE LEECH
Family *Piscicolidae*

ROUNDWORMS
Phylum *Nematoda*

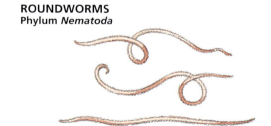

TERRESTRIAL TAPEWORM
Class *Cestoidea*

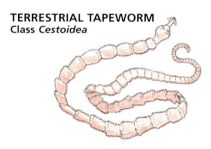

EARTHWORMS
Class *Oligochaeta*

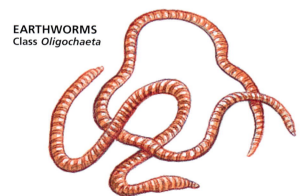

WORM EATING APPLE
While many worms are parasitic (they survive by living off their host), others eat plant or animal matter, such as fruit.

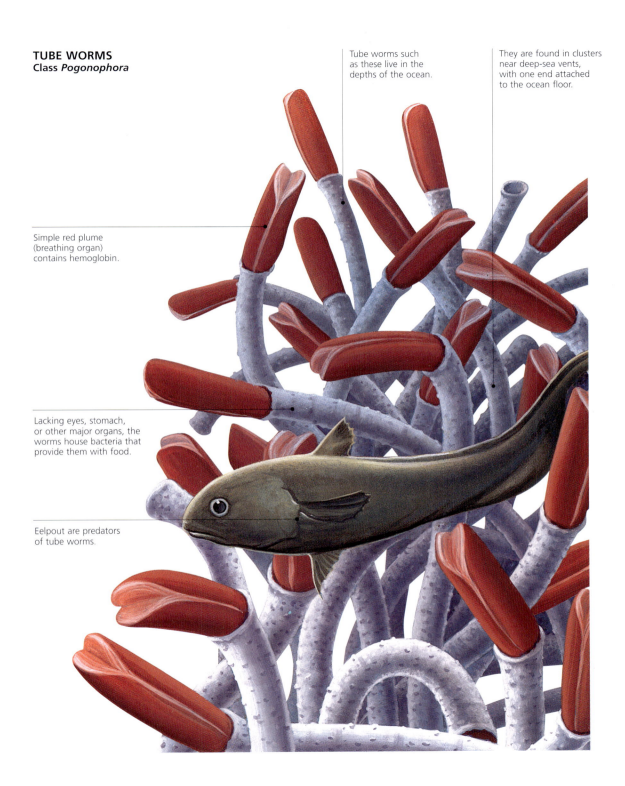

TUBE WORMS
Class *Pogonophora*

Tube worms such as these live in the depths of the ocean.

They are found in clusters near deep-sea vents, with one end attached to the ocean floor.

Simple red plume (breathing organ) contains hemoglobin.

Lacking eyes, stomach, or other major organs, the worms house bacteria that provide them with food.

Eelpout are predators of tube worms.

Echinoderms

STARFISH ANATOMY

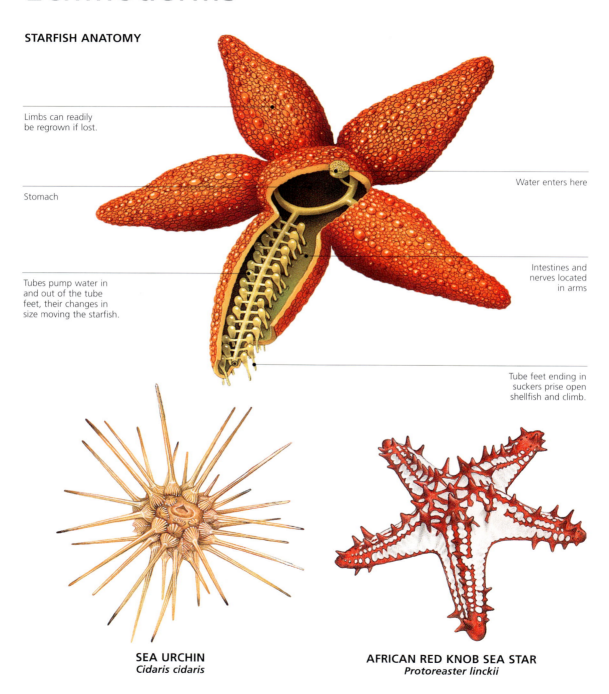

Limbs can readily be regrown if lost.

Stomach

Tubes pump water in and out of the tube feet, their changes in size moving the starfish.

Water enters here

Intestines and nerves located in arms

Tube feet ending in suckers prise open shellfish and climb.

SEA URCHIN
Cidaris cidaris

AFRICAN RED KNOB SEA STAR
Protoreaster linckii

SEA CUCUMBER
Stichopus chloronotus

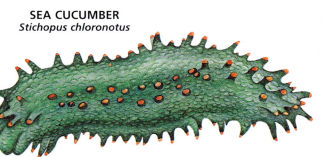

CROWN-OF-THORNS STARFISH
Acanthaster planci

Nocturnal crown-of-thorns starfish feed on coral polyps when corals are most active.

Individual starfish can devour over 10 square miles (26 sq km) annually.

Sea star	Starfish	*Baculogypsina*
		 |//

Calcarina	Starfish	Sea urchin

Arthropods

GREAT SURVIVORS
Arthropods are invertebrates that are protected by jointed exoskeletons. They are the largest and most successful group of living organisms on the planet (more than a million species), and include insects, spiders and crustaceans.

- Butterfly
- Mantid
- Cockchafer
- Wasp
- Ladybug
- Mosquito
- Cicada
- Grasshopper

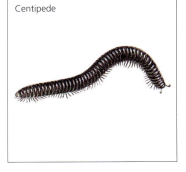

Arachnids

EVOLUTION OF ARACHNIDS

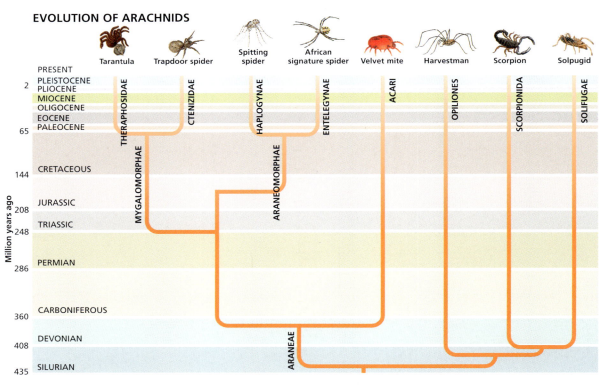

ARACHNID ANATOMY

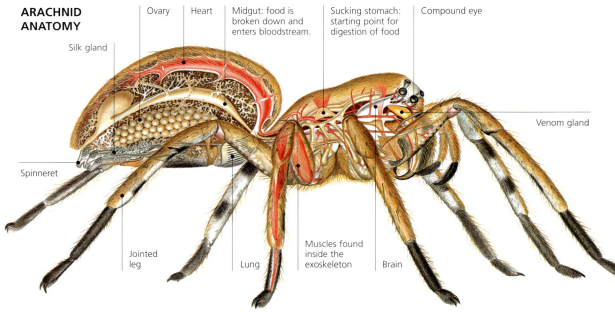

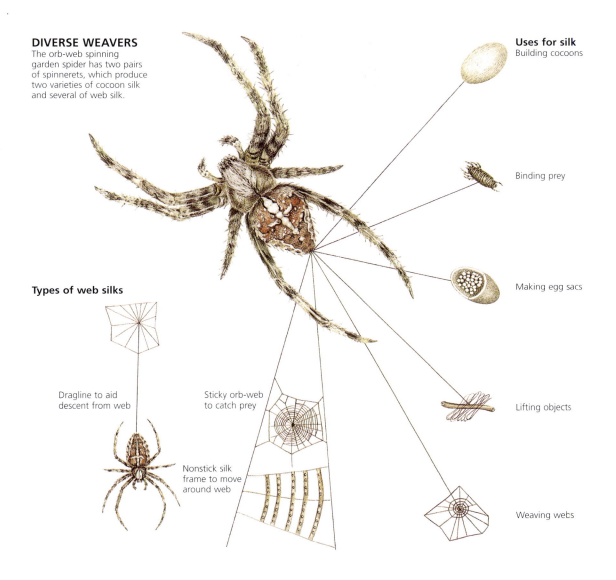

DIVERSE WEAVERS
The orb-web spinning garden spider has two pairs of spinnerets, which produce two varieties of cocoon silk and several of web silk.

Uses for silk
Building cocoons

Binding prey

Making egg sacs

Lifting objects

Weaving webs

Types of web silks

Dragline to aid descent from web

Sticky orb-web to catch prey

Nonstick silk frame to move around web

SPIDER FACES

Crab spider

Woodlouse-eating spider

Ogre-faced spider

Huntsman spider

Spiders

WHITE LADY SPIDER
Family *Sparassidae*

PRIMITIVE SPIDER
Suborder *Labidognatha*

COMB-FOOTED SPIDER
Family *Theridiidae*

JUMPING SPIDER
Family *Salticidae*

TARANTULA
Family *Theraphosidae*

LYNX SPIDER
Family *Oxyopidae*

CRAB SPIDER
Family *Thomisidae*

COURTSHIP DISPLAYS

Wolf spiders

Jumping spiders

SPIDER JAWS

Mygalomorphic (hairy) spider
Fangs point downward

Arabeomorphic (typical) spider
Fangs hinge together sideways

REPRODUCTION RITUALS

Nursery web spiders
Male presents wrapped prey to female, before mating.

Crab spiders
Male fastens female to ground with silk, then deposits sperm.

SHEDDING THE EXOSKELETON

Hangs from web

Exoskeleton tears apart

Draws fragile legs out

Waits for exoskeleton to dry

Spiders

REDBACK SPIDER
Family *Theridiidae*

BOLAS SPIDER
Family *Araneidae*

NORTHERN BLACK WIDOW
Family *Theridiidae*

SPITTING SPIDER
Family *Scytodidae*

JUMPING SPIDERS
Family *Salticidae*

MOUSE SPIDER
Family *Theraphosidae*

TRAPDOOR SPIDER
Family *Ctenizidae*

FUNNEL-WEB SPIDER
Family *Agelenidae*

VIOLIN SPIDER
Family *Loxoscelidae*

FROM EGG TO ADULT

Huntsman egg sac with eggs	Huntsman spiderling	Juvenile huntsman	Adult huntsman

Arthropods

INVERTEBRATES

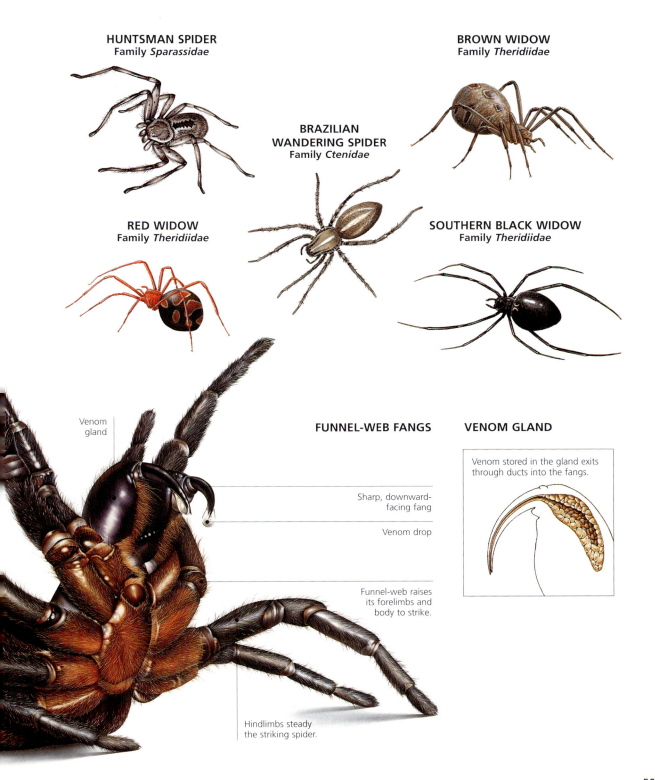

Orb-weaving Spiders

AFRICAN SIGNATURE SPIDER
Family *Araneidae*

BUILDING WEBS

Lace-sheet web

Triangle web

Hammock web

Scaffold web

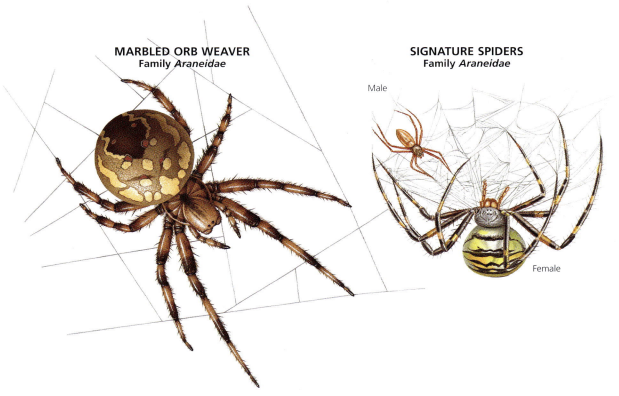

MARBLED ORB WEAVER
Family *Araneidae*

SIGNATURE SPIDERS
Family *Araneidae*

Male

Female

GOLDEN SILK SPIDER
Nephila sp.

SPINY ORB WEAVER
Family *Araneidae*

CURVED SPINY SPIDER
Gasteracantha sp.

LONG-JAWED ORB WEAVER
Family *Araneidae*

Scorpions, Mites and Ticks

HARVESTMAN
Family *Leiobunidae*

SOLPUGID
Family *Solpugidae*

HARD TICK
Family *Ixodidae*

SCORPION
Family *Buthidae*

VELVET MITE
Family *Trombidiidae*

SCORPION ANATOMY

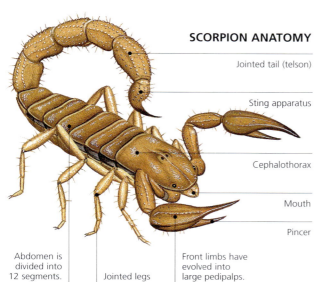

- Jointed tail (telson)
- Sting apparatus
- Cephalothorax
- Mouth
- Pincer
- Abdomen is divided into 12 segments.
- Jointed legs
- Front limbs have evolved into large pedipalps.

STING IN THE TAIL

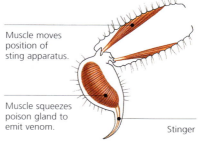

- Muscle moves position of sting apparatus.
- Muscle squeezes poison gland to emit venom.
- Stinger

SCORPION COURTSHIP DISPLAY

Some species of scorpions perform elaborate courtship displays, which involve joining pincers and 'dancing' around each other before mating.

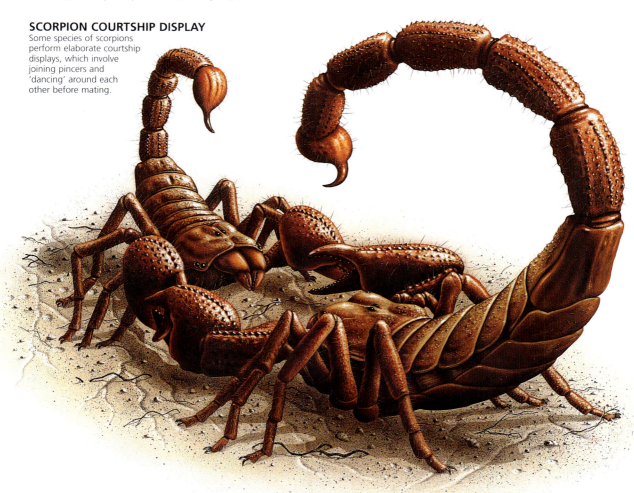

Crustaceans, Silverfishes, Centipedes and Millipedes

FLATTENED MILLIPEDE
Polydesmus complanatus

GIANT DESERT CENTIPEDE
Scolopendra heros

SPINY LOBSTER
Palinurus vulgaris

SILVERFISH
Family *Lepismatidae*

Lobster
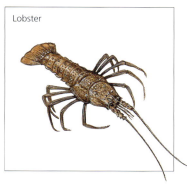

Shrimp

Centipede

Crab

SPINY LOBSTER MIGRATION

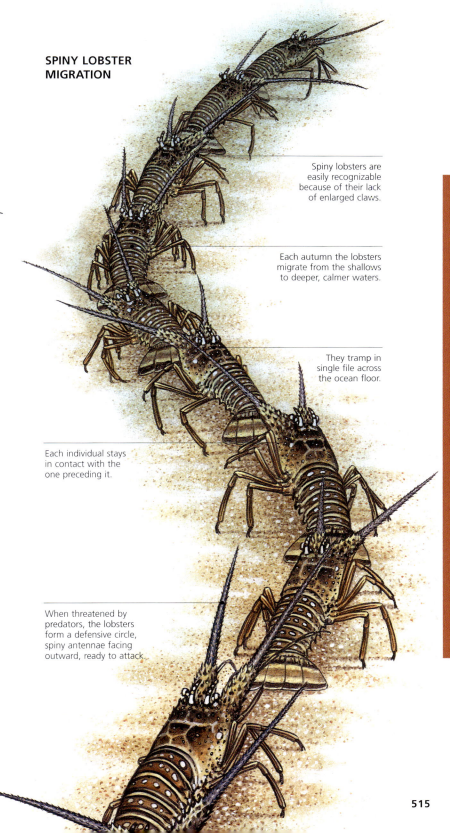

Spiny lobsters are easily recognizable because of their lack of enlarged claws.

Each autumn the lobsters migrate from the shallows to deeper, calmer waters.

They tramp in single file across the ocean floor.

Each individual stays in contact with the one preceding it.

When threatened by predators, the lobsters form a defensive circle, spiny antennae facing outward, ready to attack.

Crustaceans, Silverfishes, Centipedes and Millipedes

KRILL
Family *Euphausiidae*

MILLIPEDE
Class *Diplopoda*

SHRIMPS
Order *Decapoda*

BANDED CORAL SHRIMP
Stenopus hispidus

PURPLE SHORE CRAB
Hemigrapsus nudus

HERMIT CRAB
Family *Diogenidae*

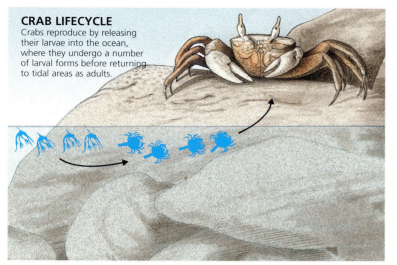

CRAB LIFECYCLE
Crabs reproduce by releasing their larvae into the ocean, where they undergo a number of larval forms before returning to tidal areas as adults.

Zoea larva

Megalopa larva

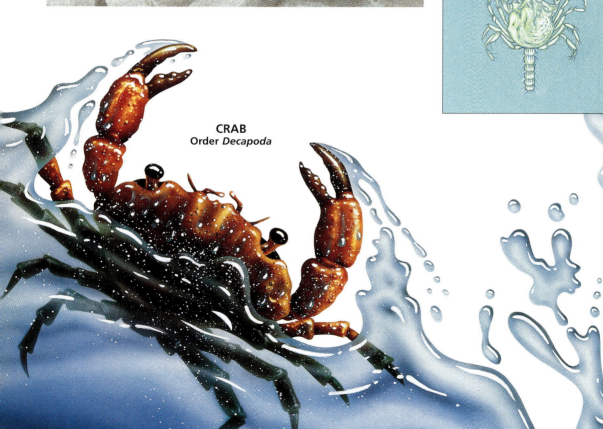

CRAB
Order *Decapoda*

Insects

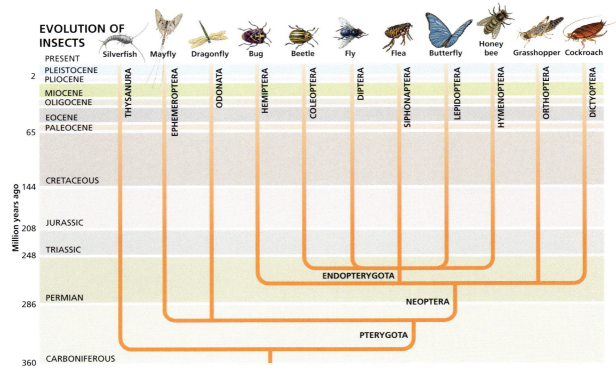

TYPES OF WINGS

True fly

Mantid

Butterfly

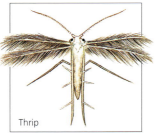
Thrip

Wasp

Dragonfly

INSECT ANATOMY

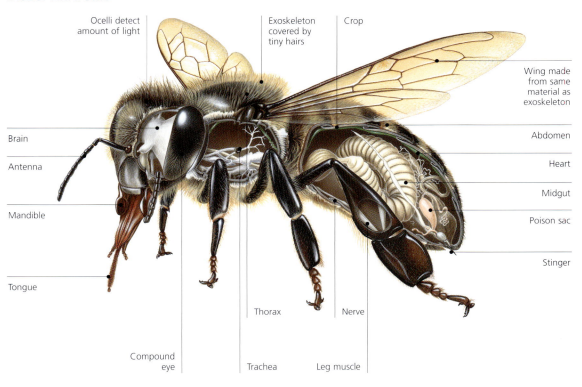

TYPES OF ANTENNAE

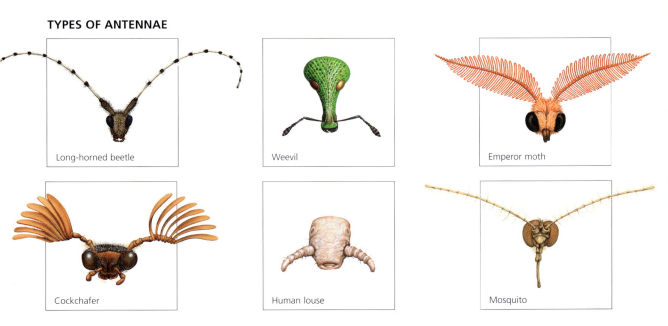

Dragonflies, Mayflies and Mantids

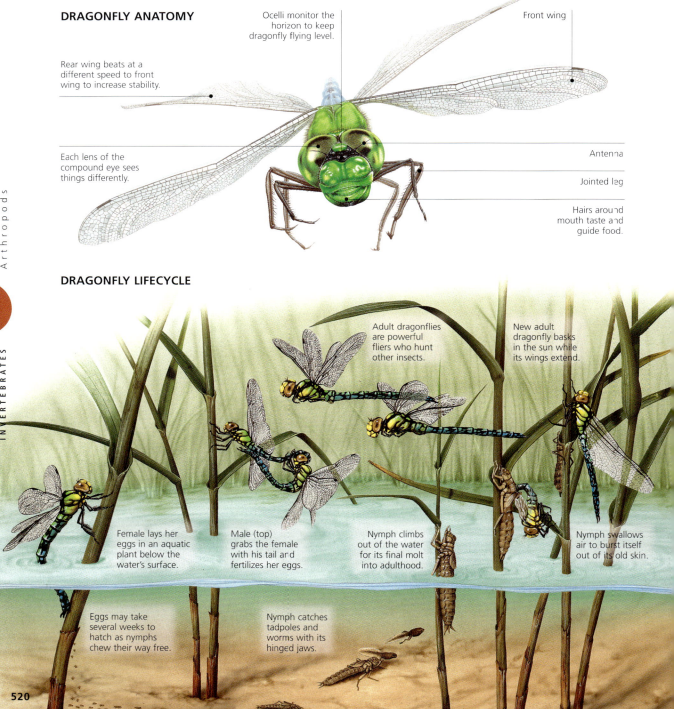

Praying mantis forelimb

Dragonfly

Dragonflies in flight

Mantis

Damselfly nymph

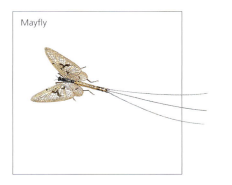
Mayfly

PRAYING MANTIS
Family *Mantidae*

ORCHID MANTIS
Family *Hymenopodidae*

Cockroaches, Termites and Lice

DAMP-WOOD TERMITE
Family *Rhinotermitidae*

FEMALE COCKROACH LAYING EGGS
As the eggs of many species of cockroaches are laid on the ground, they are covered in hard cases for protection.

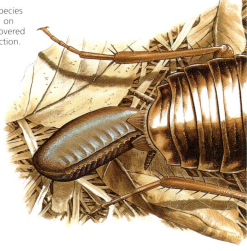

Cockroach

Termite

Spinifex termite

Cockroach head displaying antennae

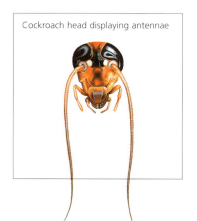

Louse

Fish louse

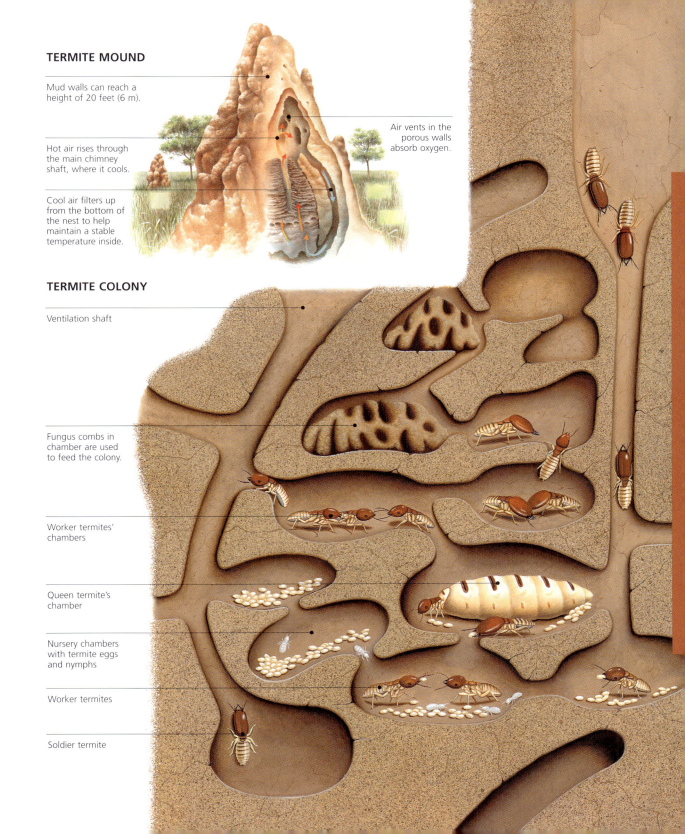

Crickets and Grasshoppers

GRASSHOPPER CALLING
Grasshoppers call by scraping pegs on their hindlegs against their wings; this causes them to 'stridulate' or vibrate.

GREAT GREEN BUSH CRICKET
Family *Tettigoniidae*

LOCUST SWARM
Locusts thrive in warm weather, forming swarms that can reach as many as 40,000 million individuals.

FIELD CRICKET
Family *Gryllidae*

JERUSALEM CRICKET
Family *Gryllacrididae*

LOCUST
Family *Acrididae*

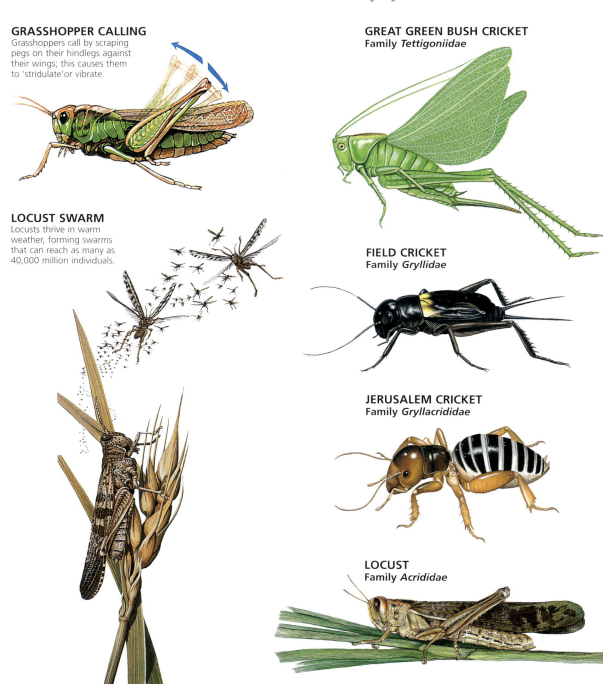

Male mole cricket in tunnel

Bush cricket

Koringkriek cricket leg

True katydid on leaf

Grasshopper spreading wings

Grasshopper

Adult earwig and nymphs feeding on branch

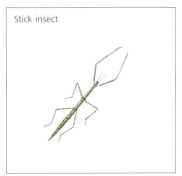
Stick insect

Grasshopper head

Nymph (top) and adult earwigs

Grasshopper leg

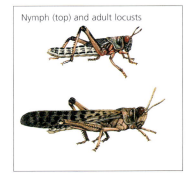
Nymph (top) and adult locusts

Bugs, Lacewings and Thrips

CICADA
Family *Cicadidae*

Anatomy of feeding
Bugs have distinctive mouthparts—a beaklike rostrum surrounded by four sharp stylets. The stylets pierce the food source, then the rostrum carries toxic saliva through the puncture to partially digest the meal before ingesting.

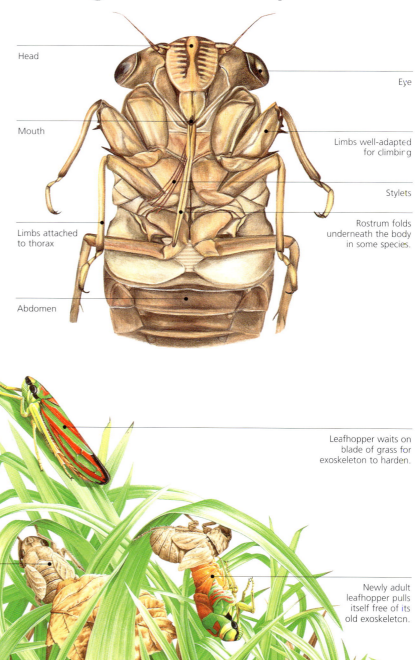

Head

Eye

Mouth

Limbs well-adapted for climbing

Stylets

Limbs attached to thorax

Rostrum folds underneath the body in some species.

Abdomen

LEAFHOPPER LIFECYCLE

Leafhopper waits on blade of grass for exoskeleton to harden.

Leafhopper nymph prepares to molt its final juvenile exoskeleton.

Newly adult leafhopper pulls itself free of its old exoskeleton.

Lacewing at rest

Lacewing taking off

Lacewing in flight

Stink bug

True bug with bright warning coloring

Milkweed bug

Lacewing

Thrip

Water bug leg

Nymph (top) and adult elder bugs

Periodical cicada

Assassin bug head

Bugs, Lacewings and Thrips

PEANUT BUGS
Fulgora sp.

Out of sight
Their dull coloration mimics the patterns of the trees they rest on, camouflaging them from hungry birds.

Spotted
If spied, a peanut bug snaps open its wings to reveal two large, eyelike spots to try to startle the would-be predator away.

Water scorpion attacks a tadpole while hanging from pond weeds.

JESTER BUG
Family *Heteroptera*

IN AND OUT OF WATER

Water boatman swims through the pond, using its legs as paddles.

Water striders use surface tension to 'walk' on the water's surface.

Many species of bugs spend larval or juvenile stages in the water.

Beetles

HERCULES BEETLES
Family *Dynastinae*
Rival male Hercules beetles grapple with their swordlike horns, battling for the right to mate with the female (right).

Diving beetle leg

Whirligig beetle swimming

Fire beetle

Diving beetle underwater

Darkling beetle

Jewel beetle

Colorado beetle

Flea beetle

Hercules beetle

Colorado beetle on branch

Large chafer

Rhinoceros beetle

Beetles

PAPUA NEW GUINEAN WEEVIL
Eupholus bennetti

TORTOISE BEETLE
Family *Chrysomelidae*

GIRAFFE WEEVIL
Family *Brentidae*

HARLEQUIN BEETLE
Family *Cerambycidae*

LONG-HORNED BEETLE
Family *Cerambycidae*

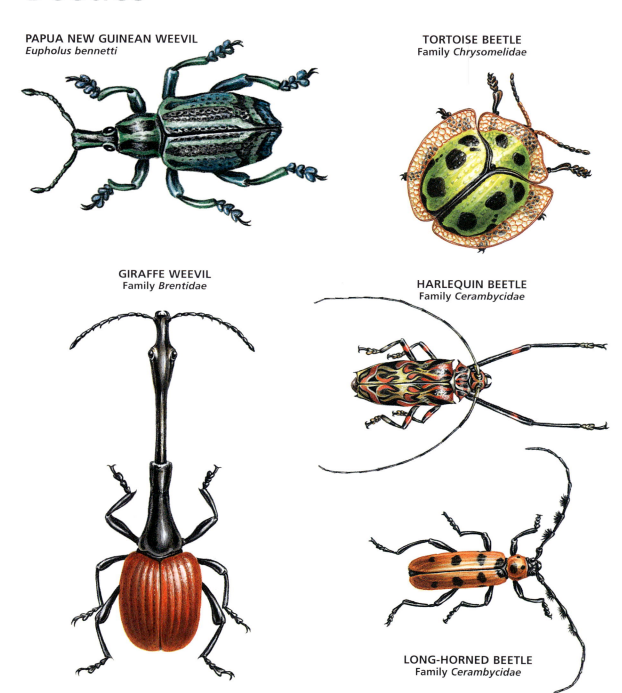

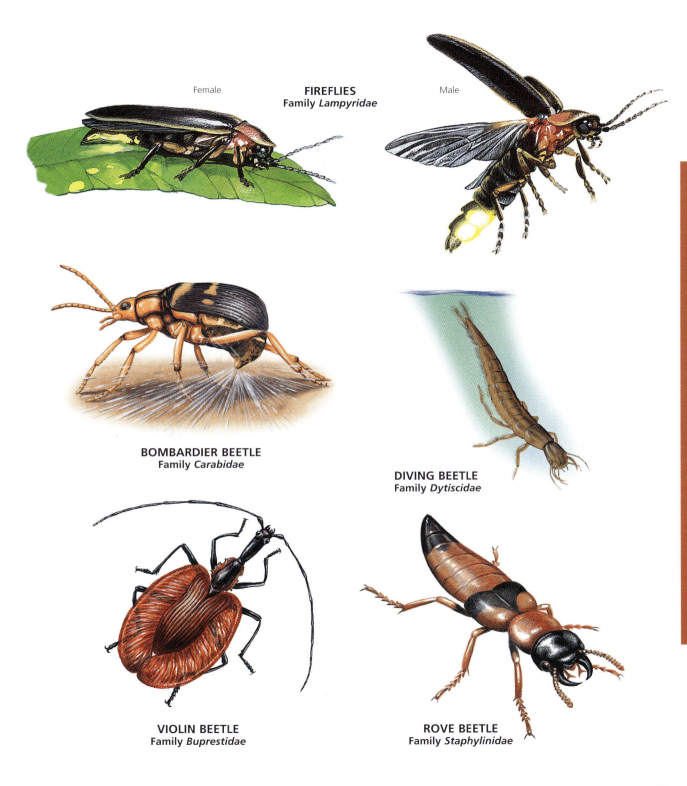

Ladybugs

TEN-SPOTTED LADYBUG
Family *Coccinellidae*

FIVE-SPOTTED LADYBUG
Family *Coccinellidae*

LADYBUG LAYING EGGS
The bright yellow eggs of the ladybug are laid on a leaf in clutches. The female chooses to lay her eggs close to a food source, such as an aphid colony.

SEVEN-SPOTTED LADYBUG LARVA
Family *Coccinellidae*

LADYBUG LIFE STAGES

Egg
Hatches at 3–5 days

Larva
Pupates at 2–3 weeks

Pupa
Changes after 7–10 days

Adult
Lives for more than one year

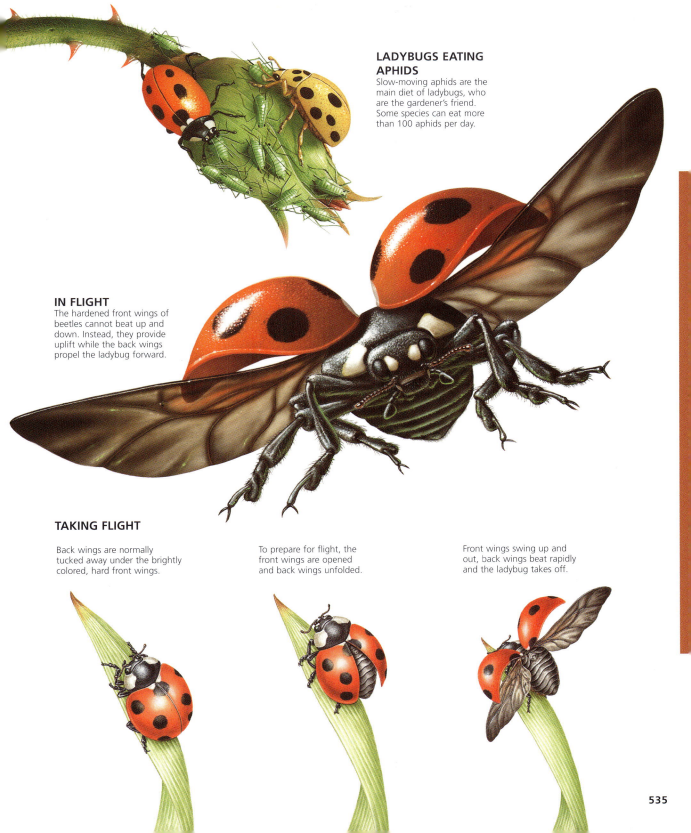

LADYBUGS EATING APHIDS
Slow-moving aphids are the main diet of ladybugs, who are the gardener's friend. Some species can eat more than 100 aphids per day.

IN FLIGHT
The hardened front wings of beetles cannot beat up and down. Instead, they provide uplift while the back wings propel the ladybug forward.

TAKING FLIGHT

Back wings are normally tucked away under the brightly colored, hard front wings.

To prepare for flight, the front wings are opened and back wings unfolded.

Front wings swing up and out, back wings beat rapidly and the ladybug takes off.

Flies, Fleas and Mosquitoes

DEER FLY
Family *Tabanidae*

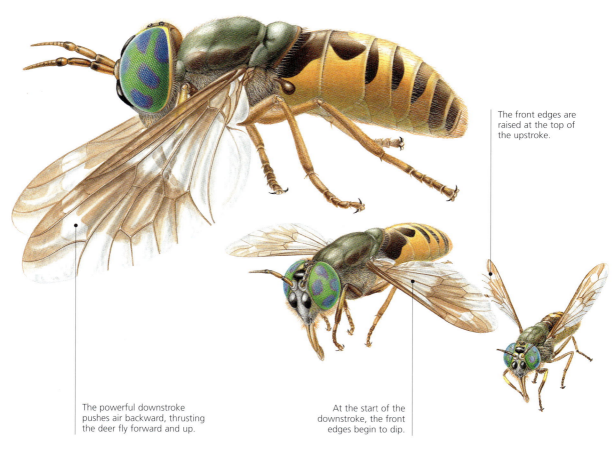

The front edges are raised at the top of the upstroke.

The powerful downstroke pushes air backward, thrusting the deer fly forward and up.

At the start of the downstroke, the front edges begin to dip.

CADDIS FLY LIFE STAGES

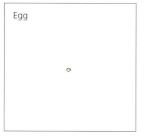

Egg

Larva

Pupa

Adult

Fly leg with sticky foot pads 	Horsefly 	Fly 
Green bottle fly 	House fly head 	Hover fly 
Blowfly maggots 	Mosquito head 	Midge 
Horsefly head 	Caddis fly 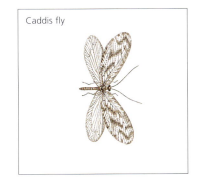	Flea jumping 

Flies, Fleas and Mosquitoes

ANOPHELES MOSQUITO
Family *Culicidae*

COMMON FLEA
Family *Pulicidae*

HOUSE FLY
Family *Muscidae*

Mosquito head: long proboscis is inserted into prey to suck blood.

Flea head: backward facing cheek combs identify species.

Fly head: colorfully patterned compound eyes aid navigation.

MOSQUITO LIFE STAGES

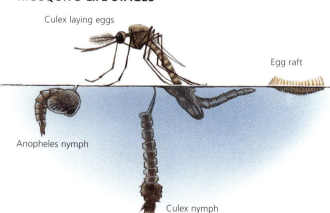

Culex laying eggs

Egg raft

Anopheles nymph

Culex nymph

CULEX MOSQUITO
Culex sp.

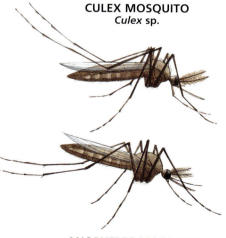

ANOPHELES MOSQUITO
Anopheles sp.

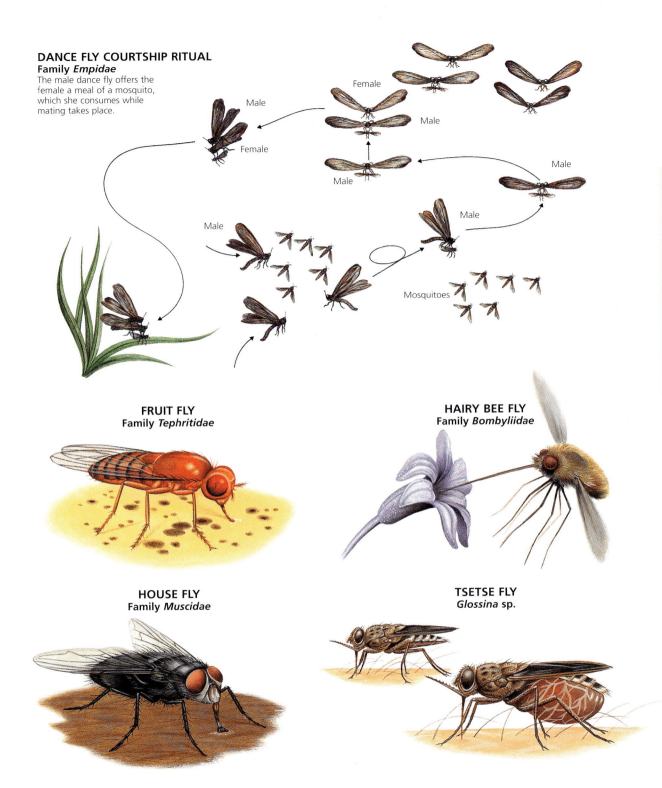

Butterflies and Moths

BUTTERFLY CHRYSALIS (PUPA)

BUTTERFLY AND MOTH COMPARISON

Butterfly wings fold vertically when resting.

Moth wings are held horizontal when resting.

Thin, threadlike antennae with clubbed tips

Feathery or straight antennae without clubbed tips

Butterfly caterpillar

Moth caterpillar

INDIAN MOON MOTH LIFECYCLE

Female Indian moon moth lays eggs on leaves.

Caterpillars (larvae) feed on nearby leaves and molt a number of times.

They then pupate (transform into pupae) inside a cocoon for several months.

A new adult moon moth emerges and hangs to strech and dry its wings.

HERCULES MOTH CATERPILLAR
Family *Saturniidae*

WOOLLY BEAR CATERPILLAR
Family *Arctiidae*

Butterflies

HACKBERRY BUTTERFLY
Family *Nymphalidae*

SWALLOWTAIL
Family *Papilionidae*

PAINTED LADY BUTTERFLY
Family *Nymphalidae*

PIPEVINE SWALLOWTAIL
Family *Papilionidae*

Anise swallowtail drinking nectar

Butterfly in flight

Butterfly at rest

DEADLY OR HARMLESS?

The harmless viceroy mimics the monarch so predators stay away.

The poisonous monarch butterfly has bold orange coloring as a warning.

88 BUTTERFLY
Diaethria sp.

MALAYAN LACEWING BUTTERFLY
Family *Nymphalidae*

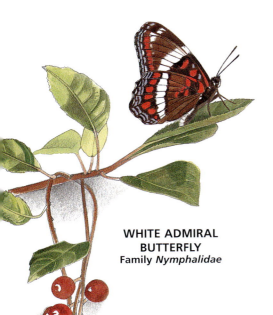

WHITE ADMIRAL BUTTERFLY
Family *Nymphalidae*

BORDERED PATCH BUTTERFLY
Family *Nymphalidae*

Moths

INDIAN MOON MOTH
Family *Saturniidae*

WHITE-LINED SPHINX MOTH
Family *Sphingidae*

HUMMINGBIRD MOTH
Family *Sphingidae*

Moth larva eating leaves

Mexican 'jumping beans' (moth larvae)

Sheep moth

FIVE-SPOTTED BURNET MOTH
Family *Zygaenidae*

VAPORER MOTH CATERPILLAR
Family *Lymantriidae*

YELLOW EMPEROR MOTH
Family *Saturniidae*

CERISY'S SPHINX MOTH
Family *Sphingidae*

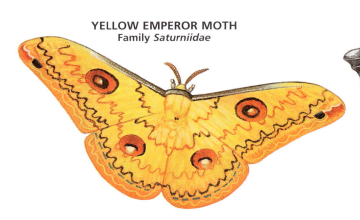

PUSS MOTH CATERPILLAR
Family *Notodontidae*

MADAGASCAN SUNSET MOTH
Family *Uraniidae*

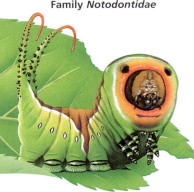

Bees, Wasps and Ants

SAND WASP
Family *Sphecidae*

BULLDOG ANT
Family *Formicidae*

BUMBLEBEE
Family *Apidae*

POTTER WASP
Family *Vespidae*

BUILDING A WASP NEST

Solitary queen begins nest.   She adds cells to ay eggs in.   First eggs hatch into workers, who continue to build cells.   Final nest may contain more than 10,000 cells.

Honey bee head

Velvet ant (species of wasp)

Army ant

Hairy-legged mining bee

Digger wasp gathering nectar

Trap-jaw ant head

ANTS AND PLANTS

Bull's horn acacia has broad, hollow thorns.

Some species of ants use them as their home.

In return for this hospitality, the ants chase off possible threats, such as beetles and cows.

QUEENLY DUTIES
Only a queen ant is able to lay eggs, which are taken by a worker and tended until adulthood.

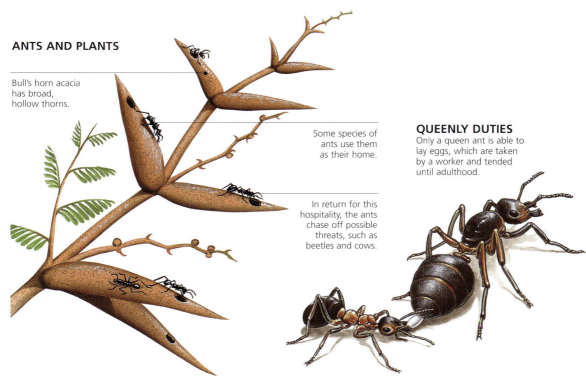

Wasps and Ants

EUROPEAN WASP
Family *Vespidae*

Venom sac

Stinger

ICHNEUMON WASP
Family *Ichneumonidae*

Female drills into a tree branch with her ovipositor (tubelike tail).

She stings a wood wasp larva, then deposits her eggs. These will feed on the larva when hatched.

TYPES OF WASPS

Social wasp

Cow killer wasp

Torymid wasp

Spider wasp

Mud dauber

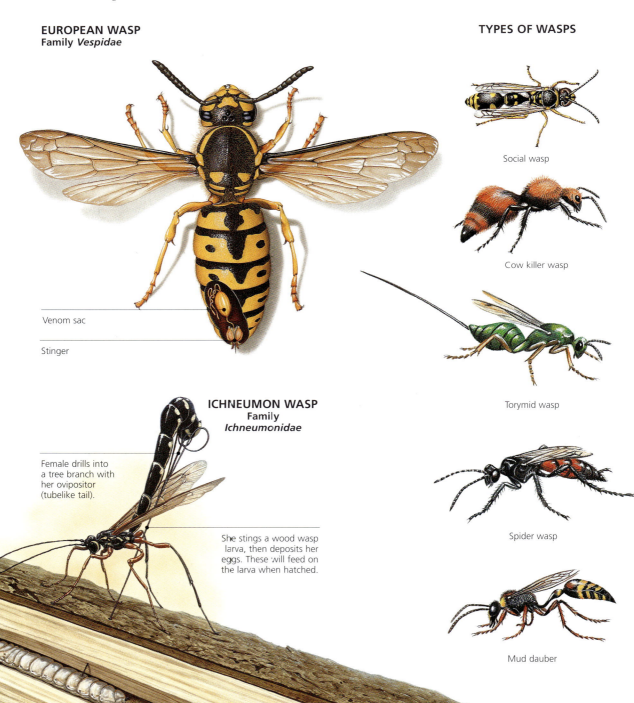

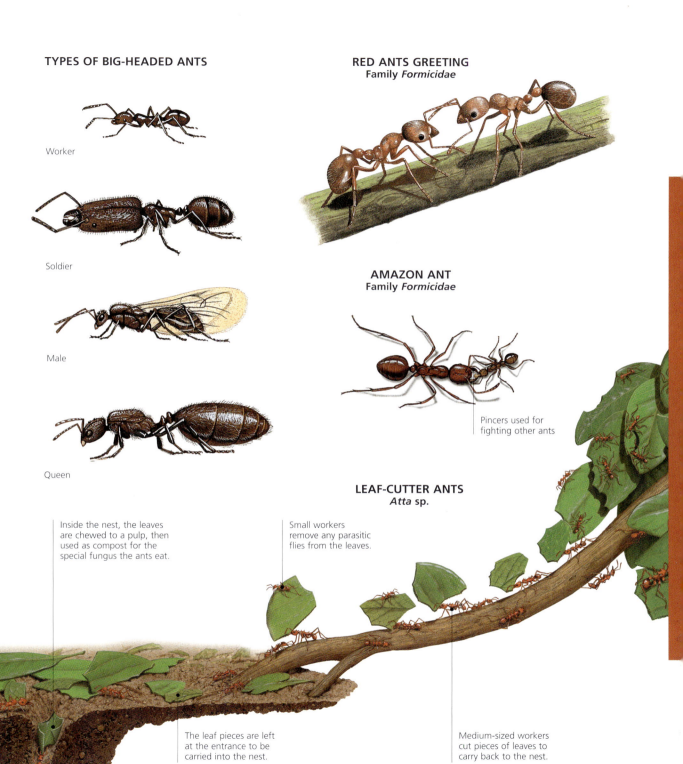

Honey Bees

HONEY BEE DANCE
Worker bees returning from a food source dance to inform others in the hive where it is.

If the food source is close by, the dance is circular, the alignment showing the bee's direction relative to the sun.

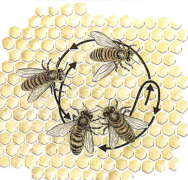

If the source is further away, the dance is a figure-of-eight, the speed and number of 'waggles' at the center indicating distance.

GATHERING NECTAR
Bees gather nectar from flowers to provide food for the hive, storing it on their legs.

Providing nectar for bees is an effective way for plants to have pollen distributed to fertilize other flowers.

After converting the nectar into honey, bees store it in cells for the hive to share.

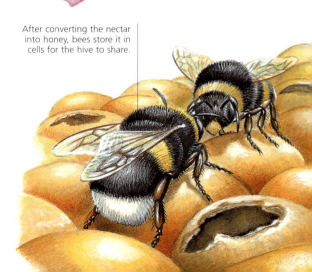

INSIDE A BEE HIVE

BEE EGG
Single egg laid in each cell

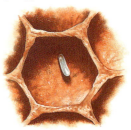

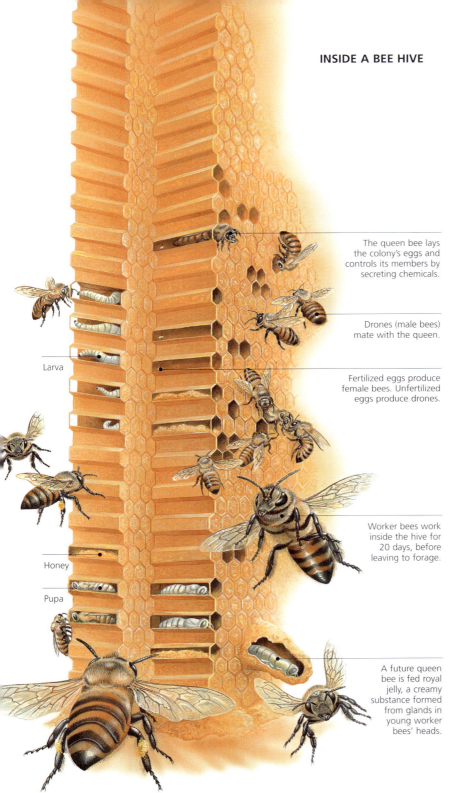

The queen bee lays the colony's eggs and controls its members by secreting chemicals.

Drones (male bees) mate with the queen.

Larva

Fertilized eggs produce female bees. Unfertilized eggs produce drones.

Worker bees work inside the hive for 20 days, before leaving to forage.

Honey

Pupa

A future queen bee is fed royal jelly, a creamy substance formed from glands in young worker bees' heads.

MOVING HIVES
Queen gathers a group of workers to leave the old hive.

The vulnerable queen is protected until a new home is found.

Scouts inform the others when a new home is found, like this beekeeper's box.

FACT FILE

Classifying Animals

LEVELS OF CLASSIFICATION

Kingdom
There are five kingdoms, of which animals is one. Plants, fungi, protists and monerans are the others.

Phylum
First subdivision within the kingdom, linking broadly similar creatures. Vertebrates (chordata) form one phylum.

Class
Subdivision within the phylum linking similar orders. Mammals, birds, reptiles, amphibians and fishes are all classes.

Order
Subdivision within the class linking family groups by common characteristics. Marsupials and bats are orders.

Family
Subdivision within the order linking closely related individuals according to biology and behavior, like vipers or geckos.

Genus
Subdivision within the family, containing one or more species that are biologically very closely related, like salmons or tunas.

Species
Individuals within the genus that can interbreed and produce offspring that can interbreed, such as ostriches.

BROWN BEAR
Ursus arctos

ANIMALIA
(Kingdom)

CHORDATA
(Phylum)

MAMMALIA
(Class)

CARNIVORA
(Order)

URSIDAE
(Family)

URSUS
(Genus)

ARCTOS
(Species)

FIRE SALAMANDER
Salamandra salamandra

ANIMALIA
(Kingdom)

CHORDATA
(Phylum)

AMPHIBIA
(Class)

CAUDATA
(Order)

SALAMANDRIDAE
(Family)

SALAMANDRA
(Genus)

SALAMANDRA
(Species)

Mammals

CLASS MAMMALIA MAMMALS

SUBCLASS PROTOTHERIA EGG-LAYING MAMMALS
ORDER MONOTREMATA MONOTREMES

Tachyglossidae	Echidnas
Ornithorhynchidae	Duck-billed platypus

SUBCLASS THERIA MODERN MAMMALS
INFRACLASS METATHERIA POUCHED MAMMALS
ORDER MARSUPIALIA MARSUPIALS

Didelphidae	American opossums

Microbiotheriidae	Colocolos
Caenolestidae	Shrew-opossums
Dasyuridae	Marsupial mice
Myrmecobiidae	Numbat
Thylacinidae	Thylacine
Notoryctidae	Marsupial mole
Peramelidae	Bandicoots
Peroryctidae	Spiny bandicoots
Vombatidae	Wombats
Phascolarctidae	Koala
Phalangeridae	Phalangers
Petauridae	Gliding phalangers
Pseudocheiridae	Ringtail possums
Burramyidae	Pygmy possums
Acrobatidae	Feathertails
Tarsipedidae	Honey possum
Macropodidae	Kangaroos, wallabies
Potoroidae	Bettongs

INFRACLASS EUTHERIA PLACENTAL MAMMALS
ORDER XENARTHRA ANTEATERS, SLOTHS AND ARMADILLOS

Myrmecophagidae	American anteaters
Bradypodidae	Three-toed sloths
Megalonychidae	Two-toed sloths
Dasypodidae	Armadillos

ORDER PHOLIDOTA PANGOLINS, SCALY ANTEATERS

Manidae	Pangolins, scaly anteater

ORDER INSECTIVORA INSECTIVORES

Solenodontidae	Solenodons
Tenrecidae	Tenrecs, otter shrews
Chrysochloridae	Golden moles
Erinaceidae	Hedgehogs, moonrats
Soricidae	Shrews
Talpidae	Moles, desmans

ORDER SCANDENTIA TREE SHREWS

Tupaiidae	Tree shrews

ORDER DERMOPTERA FLYING LEMURS (COLUGOS)

Cynocephalidae	Flying lemurs, colugos

ORDER CHIROPTERA BATS

Pteropodidae	Old World fruit bats
Rhinopomatidae	Mouse-tailed bats
Emballonuridae	Sheath-tailed bats
Craseonycteridae	Hog-nosed bat, bumblebee bat
Nycteridae	Slit-faced bats
Megadermatidae	False vampire bats
Rhinolophidae	Horseshoe bats
Hipposideridae	Old World leaf-nosed bats
Noctilionidae	Bulldog bats

Mormoopidae	Naked-backed bats
Phyllostomidae	New World leaf-nosed bats
Natalidae	Funnel-eared bats
Furipteridae	Smoky bats
Thyropteridae	Disk-winged bats
Myzopodidae	Old World sucker-footed bats
Vespertilionidae	Vespertilicnid bats
Mystacinidae	New Zealand short-tailed bats
Molossidae	Free-tailed bats

ORDER PRIMATES — PRIMATES

Cheirogaleidae	Dwarf lemurs
Lemuridae	Large lemurs
Megaladapidae	Sportive lemurs
Indridae	Leaping lemurs
Daubentoniidae	Aye-aye
Loridae	Lorises, galagos
Tarsiidae	Tarsiers
Callitrichidae	Marmosets, tamarins
Cebidae	New World monkeys
Cercopithecidae	Old World monkeys
Hylobatidae	Gibbons
Hominidae	Great apes, humans

ORDER CARNIVORA — CARNIVORES

Canidae	Dogs, foxes
Ursidae	Bears, pandas
Procyonidae	Racoons and relatives
Mustelidae	Mustelids
Viverridae	Civets and relatives
Herpestidae	Mongooses
Hyaenidae	Hyenas
Felidae	Cats
Otariidae	Sealions
Odobenidae	Walrus
Phocidae	Seals

ORDER CETACEA — WHALES, DOLPHINS AND PORPOISES

Platanistidae	River dolphins
Delphinidae	Dolphins
Phocoenidae	Porpoises
Monodontidae	Narwhal, white whale
Physeteridae	Sperm whales
Ziphiidae	Beaked whales
Eschrichtiidae	Gray whale
Balaenopteridae	Rorquals
Balaenidae	Right whales

ORDER SIRENIA — SEA COWS

Dugongidae	Dugong
Trichechidae	Manatees

ORDER PROBOSCIDEA — ELEPHANTS

Elephantidae	Elephants

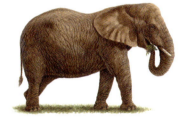

ORDER PERISSODACTYLA — ODD-TOED UNGULATES

Equidae	Horses
Tapiridae	Tapirs
Rhinocerotidae	Rhinoceroses

ORDER HYRACOIDEA — HYRAXES

Procaviidae	Hyraxes

ORDER TUBULIDENTATA — AARDVARK

Orycteropodidae	Aardvark

ORDER ARTIODACTYLA — EVEN-TOED UNGULATES

Suidae	Pigs
Tayassuidae	Peccaries
Hippopotamidae	Hippopotamuses
Camelidae	Camels, camelids
Tragulidae	Mouse deer
Moschidae	Musk deer
Cervidae	Deer

ORDER ARTIODACTYLA — EVEN-TOED UNGULATES

Giraffidae	Giraffe, okapi
Antilocapridae	Pronghorn
Bovidae	Cattle, antelopes, sheep and goats

ORDER RODENTIA — RODENTS

Aplodontidae	Mountain beaver
Sciuridae	Squirrels, marmots and relatives
Geomyidae	Pocket gophers
Heteromyidae	Pocket mice
Castoridae	Beavers
Anomaluridae	Scaly-tailed squirrels
Pedetidae	Spring hare
Muridae	Rats, mice, gerbils and relatives
Gliridae	Dormice
Seleviniidae	Desert dormouse
Zapodidae	Jumping mice
Dipodidae	Gerboas
Hystricidae	Old World porcupines
Erethizontidae	New World porcupines
Caviidae	Cavies and relatives
Hydrochaeridae	Capybara
Dinomyidae	Pacarana
Dasyproctidae	Agoutis, pacas
Chinchillidae	Chinchillas and relatives
Capromyidae	Hutias and relatives
Myocastoridae	Coypu

Octodontidae	Degus and relatives
Ctenomyidae	Tuco-tucos
Abrocomidae	Chinchilla-rats
Echymidae	Spiny rats
Thryonomyidae	Cane rats
Petromyidae	African rock-rat
Bathyergidae	African mole-rats
Ctenodactylidae	Gundis

ORDER LAGOMORPHA — LAGOMORPHS

Ochotonidae	Pikas
Leporidae	Rabbits, hares

ORDER MACROSCELIDEA — ELEPHANT SHREWS

Macroscelididae	Elephant shrews

Birds

CLASS AVES — BIRDS

ORDER STRUTHIONIFORMES — RATITES AND TINAMOUS

Struthionidae	Ostrich
Tinamidae	Tinamous
Rheidae	Rheas
Casuariidae	Cassowaries
Dromaiidae	Emu
Apterygidae	Kiwis

ORDER PROCELLARIIFORMES — ALBATROSSES AND PETRELS

Diomedeidae	Albatrosses
Procellariidae	Shearwaters
Hydrobatidae	Storm petrels
Pelecanoididae	Diving petrels

ORDER SPHENISCIFORMES — PENGUINS

Spheniscidae	Penguins

ORDER GAVIIFORMES — DIVERS

Gaviidae	Divers (loons)

ORDER PODICIPEDIFORMES — GREBES

Podicipedidae	Grebes

ORDER PELECANIFORMES — PELICANS AND RELATIVES

Phaethontidae	Tropicbirds
Pelecanidae	Pelicans
Phalacrocoracidae	Cormorants, anhingas
Sulidae	Gannets, boobies
Fregatidae	Frigatebirds

ORDER CICONIIFORMES — HERONS AND RELATIVES

Ardeidae	Herons, bitterns
Scopidae	Hammerhead
Ciconiidae	Storks
Balaenicipitidae	Whale-headed stork
Threskiornithidae	Ibises, spoonbills

ORDER PHOENICOPTERIFORMES — FLAMINGOS

Phoenicopteridae	Flamingos

ORDER FALCONIFORMES — RAPTORS

Accipitridae	Osprey, kites, hawks, eagles, Old World vultures, harriers, buzzards, harpies, buteonines
Sagittariidae	Secretarybird
Falconidae	Falcons, falconets, caracaras
Cathartidae	New World vultures

ORDER ANSERIFORMES — WATERFOWL AND SCREAMERS

Anatidae	Geese, swans, ducks
Anhimidae	Screamers

ORDER GALLIFORMES — GAMEBIRDS

Megapodiidae	Megapodes (mound-builders)
Cracidae	Chachalacas, guans, curassows
Phasianidae	Turkeys, grouse and relatives

ORDER OPISTHOCOMIFORMES — HOATZIN

Opisthocomidae	Hoatzin

ORDER GRUIFORMES — CRANES AND RELATIVES

Mesitornithidae	Mesites
Turnicidae	Hemipode-quails (button quails)
Pedionomidae	Collared hemipode
Gruidae	Cranes
Aramidae	Limpkins
Psophiidae	Trumpeters
Rallidae	Rails
Heliornithidae	Finfoots
Rhynochetidae	Kagus
Eurypygidae	Sunbittern
Cariamidae	Seriemas
Otididae	Bustards

ORDER CHARADRIIFORMES — WADERS AND SHOREBIRDS

Jacanidae	Jacanas
Rostratulidae	Painted snipe
Dromadidae	Crab plover
Haematopodidae	Oystercatchers
Ibidorhynchidae	Ibisbill
Recurvirostridae	Stilts, avocets
Burhinidae	Stone curlews (thick knees)
Glareolidae	Coursers, pratincoles

Charadriidae	Plovers, dotterels
Scolopacidae	Curlews, sandpipers, snipes
Thinocoridae	Seedsnipes
Chionididae	Sheathbills
Laridae	Gulls, terns, skimmers
Stercorariidae	Skuas, jaegers
Alcidae	Auks

ORDER COLUMBIFORMES — PIGEONS AND SANDGROUSE

Pteroclididae	Sandgrouse
Columbidae	Pigeons, doves

ORDER PSITTACIFORMES — PARROTS

Cacatuidae	Cockatoos
Psittacidae	Parrots

ORDER CUCULIFORMES — TURACOS AND CUCKOOS

Musophagidae	Turacos, louries (plantain-eaters)
Cuculidae	Cuckoos and relatives

ORDER STRIGIFORMES — OWLS

Tytonidae	Barn owls, bay owls
Strigidae	Hawk owls (true owls)

ORDER CAPRIMULGIFORMES — NIGHTJARS AND FROGMOUTHS

Steatornithidae	Oilbird
Podargidae	Frogmouths
Nyctibiidae	Potoos
Aegothelidae	Owlet nightjars
Caprimulgidae	Nightjars

ORDER APODIFORMES — SWIFTS AND HUMMINGBIRDS

Apodidae	Swifts
Hemiprocnidae	Crested swifts
Trochilidae	Hummingbirds

ORDER COLIIFORMES — MOUSEBIRDS

Coliidae	Mousebirds

ORDER TROGONIFORMES — TROGONS

Trogonidae	Trogons

ORDER CORACIIFORMES — KINGFISHERS AND RELATIVES

Alcedinidae	Kingfishers
Todidae	Todies
Momotidae	Motmots
Meropidae	Bee-eaters
Coraciidae	Rollers
Upupidae	Hoopoe
Phoeniculidae	Wood-hoopoes
Bucerotidae	Hornbills

ORDER PICIFORMES — WOODPECKERS AND BARBETS

Galbulidae	Jacamars
Bucconidae	Puffbirds
Capitonidae	Barbets
Ramphastidae	Toucans
Indicatoridae	Honeyguides
Picidae	Woodpeckers

ORDER PASSERIFORMES — PASSERINES OR PERCHING BIRDS

Suborder Eurylaimi — Broadbills and Pittas

Eurylaimidae	Broadbills
Philepittidae	Sunbirds, asitys
Pittidae	Pittas
Acanthisittidae	New Zealand wrens

Suborder Furnarii — Ovenbirds and relatives

Dendrocolaptidae	Woodcreepers
Furnariidae	Ovenbirds
Formicariidae	Antbirds
Rhinocryptidae	Tapaculos

Suborder Tyranni — Tyrant Flycatchers and Relatives

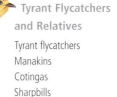

Tyrannidae	Tyrant flycatchers
Pipridae	Manakins
Cotingidae	Cotingas
Oxyruncidae	Sharpbills
Phytotomidae	Plantcutters

Suborder Oscines — Songbirds

Menuridae	Lyrebirds
Atrichornithidae	Scrub-birds
Alaudidae	Larks
Motacillidae	Wagtails, pipits
Hirundinidae	Swallows, martins
Campephagidae	Cuckoo-shrikes and relatives
Pycnonotidae	Bulbuls
Irenidae	Leafbirds, ioras, bluebirds
Laniidae	Shrikes
Vangidae	Vangas
Bombycillidae	Waxwings
Hypocoliidae	Hypocolius
Ptilogonatidae	Silky flycatchers
Dulidae	Palmchat
Prunellidae	Accentors, hedge-sparrows
Mimidae	Mockingbirds and relatives
Cinclidae	Dippers
Turdidae	Thrushes
Timaliidae	Babblers and relatives
Troglodytidae	Wrens
Sylviidae	Old World warblers
Muscicapidae	Old World flycatchers
Maluridae	Fairywrens and relatives
Acanthizidae	Australian warblers and relatives
Ephthianuridae	Australian chats
Orthonychidae	Logrunners and relatives
Rhipiduridae	Fantails
Monarchidae	Monarch flycatchers
Petroicidae	Australasian robins
Pachycephalidae	Whistlers and relatives
Aegithalidae	Long-tailed tits
Remizidae	Penduline tits

Paridae	True tits, chickadees, titmice
Sittidae	Nuthatches, sitellas, wallcreeper
Certhiidae	Holarctic treecreepers
Rhabdornithidae	Philippine treecreepers
Climacteridae	Australasian treecreepers
Dicaeidae	Flowerpeckers, pardalotes
Nectariniidae	Sunbirds
Zosteropidae	White-eyes
Meliphagidae	Honeyeaters
Vireonidae	Vireos
Emberizidae	Buntings, tanagers
Parulidae	New World wood warblers
Icteridae	Icterids (American blackbirds)
Fringillidae	Finches
Drepanididae	Hawaiian honeycreepers
Estrildidae	Estrildid finches
Ploceidae	Weavers
Passeridae	Old World sparrows
Sturnidae	Starlings, mynahs
Oriolidae	Orioles, figbirds
Dicruridae	Drongos
Callaeidae	New Zealand wattlebirds
Grallinidae	Magpie-larks
Artamidae	Wood swallows
Cracticidae	Bell magpies
Ptilonorhynchidae	Bowerbirds
Paradisaeidae	Birds-of-paradise
Corvidae	Crows, jays

Reptiles

CLASS REPTILIA — REPTILES

SUBCLASS EUREPTILIA
SUPERORDER TESTUDINE
ORDER TESTUDINATA — TURTLES, TERRAPINS AND TORTOISES

Suborder Pleurodira — Side-neck Turtles

Chelidae	Snake-neck turtles
Pelomedusidae	Helmeted side-neck turtles

Suborder Cryptodira — Hidden-necked Turtles

Superfamily Trionychoidea

Kinosternidae	Mud and musk turtles
Dermatemydidae	Mesoamerican river turtle
Carretochelyidae	Australian softshell turtle
Trionychidae	Holarctic and paleotropical softshell turtles

Superfamily Cheloniodea

Dermochelyidae	Leatherback sea turtles
Cheloniidae	Sea turtles

Superfamily Testudinoidea

Chelydridae	Snapping turtles
Emydidae	New World pond turtles and terrapins
Testudinidae	Tortoises
Bataguridae	Old World pond turtles

SUPERORDER ARCHOSAURIA
ORDER CROCODILIA — CROCODILIANS

Alligatoridae	Alligators and caimans
Crocodylidae	Crocodiles
Gavialidae	Gharials

SUPERORDER LEPIDOSAURIA

ORDER RHYNCHOCEPHALIA	TUATARAS
Sphenodontidae	Tuataras

ORDER SQUAMATA	SQUAMATES
Suborder Amphisbaenia	**Amphisbaenians**
Bipedidae	Ajolotes
Amphisbaenidae	Worm lizards
Trogonophidae	Desert ringed lizards
Rhineuridae	Florida worm lizard

Suborder Iguania — **Iguanid Lizards**

Corytophanidae	Helmeted lizards
Crotaphytidae	Collared and leopard lizards
Hoplocercidae	Hoplocercids
Iguanidae	Iguanas
Opluridae	Madagascar iguanians
Phrynosomatidae	Scaly, sand and horned lizards
Polychrotidae	Anoloid lizards
Tropiduridae	Tropidurids
Agamidae	Agamid lizards
Chameleonidae	Chameleons

Suborder Scleroglossa

Superfamily Gekkonoidea

Eublepharidae	Eye-lash geckos
Gekkonidae	Geckos
Pygopodidae	Australasian flapfoots

Superfamily Scincoidea

Xantusiidae	Night lizards
Lacertidae	Lacertids
Scincidae	Skinks
Dibamidae	Dibamids
Cordylidae	Girdle-tailed lizards
Gerrhosauridae	Plated lizards
Teiidae	Macroteiids
Gymnophthalmidae	Microteiids

Superfamily Anguoidea

Xenosauridae	Knob-scaled lizards
Anguidae	Anguids; glass and alligator lizards
Helodermatidae	Beaded lizards
Varanidae	Monitor lizards
Lanthanotidae	Earless monitor lizard

Suborder Serpentes — **Snakes**

Infraorder Scolecophidia

Anomalepididae	Blind wormsnakes
Typhlopidae	Blind snakes
Leptotyphlopidae	Thread snakes

Infraorder Alethinophidia

Anomochelidae	Stump heads
Aniliidae	Coral pipesnakes
Cylindrophidae	Asian pipesnakes
Uropeltidae	Shield tails
Xenopeltidae	Sunbeam snake
Loxocemidae	Dwarf boa
Boidae	Pythons and boas

Ungaliophiidae	Ungaliophiids
Bolyeriidae	Round Island snakes
Tropidophiidae	Woodsnakes
Acrochordidae	File snakes
Atractaspididae	Mole vipers
Colubridae	Harmless and rear-ranged snakes
Elapidae	Cobras, kraits, coral snakes and sea snakes
Viperidae	Adders and vipers

Amphibians

CLASS AMPHIBIA — AMPHIBIANS

SUBCLASS LISSAMPHIBIA

ORDER CAUDATA — SALAMANDERS AND NEWTS

Suborder Sirenoidea

Sirenidae	Sirens

Suborder Cryptobranchoidea

Cryptobranchidae	Hellbenders and giant salamanders
Hynobiidae	Hynobiids

Suborder Salamandroidea

Amphiumidae	Amphiumas (congo eels)
Plethodontidae	Lungless salamanders
Rhyacotritonidae	Torrent salamanders
Proteidae	Mudpuppies, waterdogs, and the olm
Salamandridae	Salamandrids

Ambystomatidae	Mole salamanders
Dicamptodontidae	Dicamptodontids

ORDER GYMNOPHIONA — CAECILIANS

Rhinatrematidae	South American tailed caecilians
Ichthyophiidae	Ichthyophiids
Uraeotyphlidae	Uraeotyphlids
Scolecomorphidae	Scolecomorphids
Caeciliidae	Caeciliids and aquatic caecilians

ORDER ANURA — FROGS AND TOADS

Ascaphidae	"Tailed" frogs
Leiopelmatidae	New Zealand frogs
Bombinatoridae	Fire-bellied toads and allies
Discoglossidae	Discoglossid frogs
Pipidae	Pipas and "clawed" frogs
Rhinophrynidae	Cone-nosed frog
Megophryidae	Megophryids
Pelodytidae	Parsley frogs
Pelobatidae	Spadefoots

Superfamily Bufonoidea

Allophrynidae	Allophrynid frog
Brachycephalidae	Saddleback frogs
Bufonidae	Toads
Helophrynidae	Ghost frogs
Leptodactylidae	Neotropical frogs
Myobatrachidae	Australasian frogs
Sooglossiidae	Seychelles frogs
Rhinodermatidae	Darwin's frogs
Hylidae	Hylid treefrogs
Pelodryadidae	Australasian treefrogs
Centrolenidae	Glass frogs
Pseudidae	Natator frogs
Dendrobatidae	Dart-poison frogs

Superfamily Ranoidea

Microhylidae	Microhylids
Hemisotidae	Shovel-nosed frogs
Arthroleptidae	Squeakers
Ranidae	Ranid frogs
Hyperoliidae	Reed and lily frogs
Rhacophoridae	Rhacaphorid treefrogs

Fishes

SUPERCLASS AGNATHA — JAWLESS FISHES

CLASS MYXINI — HAGFISHES

ORDER MYXINIFORMES	HAGFISHES
Myxinidae	Hagfishes

CLASS CEPHALASPIDOMORPHI — LAMPREYS AND RELATIVES

ORDER PETROMYZONTIFORMES	LAMPREYS
Petromyzontidae	Northern lampreys
Geotriidae	Pouched lampreys
Mordaciidae	Shorthead lampreys

SUPERCLASS GNATHOSTOMATA — JAWED FISHES

CLASS CHONDRICHTHYES — CARTILAGINOUS FISHES

SUBCLASS ELASMOBRANCHII — SHARKS AND RAYS

ORDER HEXANCHIFORMES	SIX- AND SEVENGILL SHARKS
Chlamydoselachidae	Frill shark
Hexanchidae	Sixgill, sevengill (cow) sharks

ORDER SQUALIFORMES	DOGFISH SHARKS AND RELATIVES
Echinorhinidae	Bramble sharks
Squalidae	Dogfish sharks
Oxynotidae	Roughsharks

ORDER PRISTIOPHORIFORMES	SAWSHARKS
Pristiophoridae	Sawsharks

ORDER HETERODONTIFORMES	HORN (BULLHEAD) SHARKS
Heterodontidae	Horn (bullhead) sharks

ORDER ORECTOLOBIFORMES	CARPETSHARKS AND RELATIVES
Parascylliidae	Collared carpetsharks
Brachaeluridae	Blind sharks
Orectolobidae	Wobbegongs
Hemiscylliidae	Longtail carpetsharks
Stegostomatidae	Zebra shark
Ginglymostomatidae	Nurse sharks
Rhincodontidae	Whale shark

ORDER LAMNIFORMES	MACKEREL SHARKS AND RELATIVES
Odontaspididae	Sand tigers (gray nurse sharks)
Mitsukurinidae	Goblin shark
Pseudocarchariidae	Crocodile shark
Megachasmidae	Megamouth shark
Alopiidae	Thresher sharks
Cetorhinidae	Basking shark
Lamnidae	Mackerel sharks (makos)

ORDER CARCHARHINIFORMES	GROUND SHARKS
Scyliorhinidae	Catsharks
Proscylliidae	Finback catsharks
Pseudotriakidae	False catshark
Leptochariidae	Barbled houndshark
Triakidae	Houndsharks
Hemigaleidae	Weasel sharks
Carcharhinidae	Requiem (whaler) sharks
Sphyrnidae	Hammerhead sharks

ORDER SQUATINIFORMES ANGEL SHARKS

Squatinidae	Angel sharks

ORDER RHINOBATIFORMES SHOVELNOSE RAYS

Platyrhinidae	Platyrhinids
Rhinobatidae	Guitarfishes
Rhynchobatidae	Sharkfin guitarfishes

ORDER RAJIFORMES SKATES

Rajidae	Skates

ORDER PRISTIFORMES SAWFISHES

Pristidae	Sawfishes

ORDER TORPEDINIFORMES ELECTRIC RAYS

Torpedinidae	Electric rays
Hypnidae	Coffinrays
Narcinidae	Narcinids
Narkidae	Narkids

ORDER MYLIOBATIFORMES STINGRAYS AND RELATIVES

Potamotrygonidae	River rays
Dasyatididae	Stingrays
Urolophidae	Round stingrays
Gymnuridae	Butterfly rays
Hexatrygonidae	Sixgill rays
Myliobatididae	Eagle rays
Rhinopteridae	Cownose rays
Mobulidae	Mantas, devil rays

SUBCLASS HOLOCEPHALI CHIMAERAS
ORDER CHIMAERIFORMES CHIMAERAS

Callorhynchidae	Plownose chimaeras (Elephant fishes)
Chimaeridae	Shortnose chimaeras (Ghost sharks, ratfishes)
Rhinochimaeridae	Longnose chimaeras (spookfishes)

CLASS OSTEICHTHYES BONY FISHES

SUBCLASS SARCOPTERYGII LUNGFISHES AND COELACANTH (FLESHY-FINNED FISHES)

ORDER CERATODONTIFORMES LUNGFISHES

Ceratodontidae	Australian lungfish

ORDER LEPIDOSIRENIFORMES LUNGFISHES

Lepidosirenidae	South American lungfish
Protopteridae	African lungfishes

ORDER COELACANTHIFORMES COELACANTH

Latimeriidae	Coelacanth (gombessa)

SUBCLASS ACTINOPTERYGII RAY-FINNED FISHES
ORDER POLYPTERIFORMES BICHIRS

Polypteridae	Bichirs

ORDER ACIPENSERIFORMES STURGEONS AND RELATIVES

Acipenseridae	Sturgeons
Polyodontidae	Paddlefishes

ORDER LEPISOSTEIFORMES GARS

Lepisosteidae	Gars

ORDER AMIIFORMES BOWFIN

Amiidae	Bowfin

GROUP TELEOSTEI — TELEOSTS

ORDER HIODONTIFORMES — MOONEYES

Hiodontidae — Mooneyes

ORDER OSTEOGLOSSIFORMES — BONYTONGUES AND RELATIVES

Osteoglossidae — Bonytongues
Pantodontidae — Freshwater butterflyfish
Notopteridae — Featherbacks
Mormyridae — Elephantfishes
Gymnarchidae — Aba

ORDER ELOPIFORMES — TARPONS AND RELATIVES

Elopidae — Tenpounders (ladyfishes)
Megalopidae — Tarpons

ORDER ALBULIFORMES — BONEFISHES AND RELATIVES

Albulidae — Bonefishes

ORDER NOTACANTHIFORMES — SPINY EELS AND RELATIVES

Halosauridae — Halosaurs
Notacanthidae — Spiny eels

ORDER ANGUILLIFORMES — EELS

Anguillidae — Freshwater eels
Heterenchelyidae — Shortfaced eels
Moringuidae — Spaghetti eels
Chlopsidae — False moray eels
Myrocongridae — Myrocongrid
Muraenidae — Moray eels
Nemichthyidae — Snipe eels
Muraenesocidae — Pike eels
Synaphobranchidae — Cutthroat eels
Ophichthidae — Snake eels, worm eels
Nettastomatidae — Duckbill eels
Colocongridae — Colocongrids
Congridae — Conger eels, garden eels
Derichthyidae — Narrowneck eels, spoonbill eels
Serrivomeridae — Sawtooth eels
Cyematidae — Bobtail snipe eels

Saccopharyngidae — Swallowers
Eurypharyngidae — Gulper eels
Monognathidae — Singlejaw eels

ORDER CLUPEIFORMES — SARDINES, HERRINGS, AND ANCHOVIES

Denticipitidae — Denticle herring
Clupeidae — Sardines, herrings, pilchards
Engraulididae — Anchovies
Chirocentridae — Wolf herrings

ORDER GONORYNCHIFORMES — MILKFISH, BEAKED SALMONS AND RELATIVES

Chanidae — Milkfish
Gonorynchidae — Beaked salmons
Kneriidae — Shellears
Phractolaemidae — Hingemouth

ORDER CYPRINIFORMES — CARPS, MINNOWS AND RELATIVES

Cyprinidae — Carps, minnows
Psilorhynchidae — Psilorhynchids
Balitoridae — Hillstream loaches
Cobitidae — Loaches
Gyrinocheilidae — Algae eaters
Catostomidae — Suckers

ORDER CHARACIFORMES — CHARACINS AND RELATIVES

Citharinidae — Citharinids
Distichodontidae — Distichodontins, African fin eaters
Hepsetidae — African pike characin
Erythrinidae — Trahiras and relatives
Ctenoluciidae — American pike characins
Lebiasinidae — Voladoras, pencil fishes
Characidae — African tetras (characins), characins, tetras, brycons, piranhas and relatives
Curimatidae — Toothless characins
Prochilodontidae — Flannelmouth characins
Anostomidae — Anostomins, leporinins
Hemiodontidae — Hemiodontids
Chilodontidae — Headstanders
Gasteropelecidae — Freshwater hatchetfishes

ORDER SILURIFORMES	CATFISHES AND KNIFEFISHES
Suborder Siluroidei	
Diplomystidae	Velvet catfishes
Nematogenyidae	Mountain catfish
Trichomycteridae	Spinyhead catfishes, candirus
Callichthyidae	Plated catfishes
Scoloplacidae	Spinynose catfishes
Astroblepidae	Andes catfishes
Loricariidae	Armored suckermouth catfishes
Ictaluridae	Bullhead catfishes, madtoms
Bagridae	Bagrid catfishes
Cranoglanididae	Armorhead catfishes
Siluridae	Sheatfishes, wels, glass catfishes
Schilbidae	Schilbid catfishes
Pangasiidae	Pangasiid catfishes
Amblycipitidae	Torrent catfishes
Amphiliidae	Loach catfishes
Akysidae	Asian banjo catfishes
Sisoridae	Hillstream catfishes
Clariidae	Labyrinth (airbeathing) catfishes
Heteropneustidae	Airsac catfishes
Chacidae	Square head (angler) catfishes
Olyridae	Olyrid catfishes
Malapteruridae	Electric catfishes
Ariidae	Sea catfishes
Plotosidae	Eeltail catfishes
Mochokidae	Upsidedown catfishes (squeakers)
Doradidae	Thorny catfishes
Auchenipteridae	Wood catfishes
Pimelodidae	Longwhisker, shovelnose catfishes
Ageneiosidae	Bottlenose (barbelless) catfishes
Helogenidae	Helogene catfishes
Cetopsidae	Carneros, shark (whale) catfishes
Hypophthalmidae	Loweye catfishes
Aspredinidae	Banjo catfishes
Suborder Gymnotoidei	
Sternopygidae	Longtail knifefishes
Rhamphichthyidae	Longsnout knifefishes
Hypopomidae	Bluntnose knifefishes
Apternotidae	Black knifefishes
Gymnotidae	Banded knifefishes
Electrophoridae	Electric eel (electric knifefish)

ORDER ARGENTINIFORMES	HERRINGS, SMELTS, BARRELEYES AND RELATIVES
Argentinidae	Herring smelts
Microstomatidae	Deepsea smelts
Opisthoproctidae	Barreleyes (spookfishes)
Alepocephalidae	Slickheads
Bathylaconidae	Bathylaconids
Platytroctidae	Tubeshoulders

ORDER SALMONIFORMES	SALMONS, SMELTS AND RELATIVES
Osmeridae	Northern smelts, ayu
Salangidae	Icefishes (noodlefishes)
Retropinnidae	Southern smelts
Galaxiidae	Galaxiids
Lepidogalaxiidae	Salamanderfish
Salmonidae	Trouts, salmons, chars, whitefishes

ORDER ESOCIFORMES	PIKES, PICKERELS AND RELATIVES
Esocidae	Pikes
Umbridae	Mudminnows

ORDER ATELEOPODIFORMES	JELLYNOSE FISHES
Ateleopodidae	Jellynose fishes

ORDER STOMIIFORMES	DRAGONFISHES, LIGHTFISHES AND RELATIVES
Gonostomatidae	Lightfishes, bristlemouths
Sternoptychidae	Hatchetfishes and relatives
Phosichthyidae	Lightfishes
Stomiidae	Viperfishes, dragonfishes, snaggletooths, loosejaws

ORDER AULOPIFORMES	LIZARDFISHES AND RELATIVES
Aulopidae	Aulopids
Bathysauridae	Bathysaurids
Chlorophthalmidae	Greeneyes
Ipnopidae	Ipnopids, tripod fishes
Synodontidae	Lizardfishes, Bombay duck
Scopelarchidae	Pearleyes
Notosudidae	Waryfishes
Giganturidae	Telescope fishes
Paralepididae	Barracudinas
Anotopteridae	Daggertooth
Evermannellidae	Sabertooth fishes
Omosudidae	Omosudid
Alepisauridae	Lancetfishes
Pseudotrichonotidae	Pseudotrichonotids

ORDER MYCTOPHIFORMES	LANTERNFISHES AND RELATIVES
Neoscopelidae	Neoscopelids
Myctophidae	Lanternfishes

ORDER LAMPRIDIFORMES	OARFISHES AND RELATIVES
Lampridae	Opahs
Veliferidae	Velifers
Lophotidae	Crestfishes
Radiicephalidae	Inkfishes
Trachipteridae	Ribbonfishes
Regalecidae	Oarfishes
Stylephoridae	Tube eye

ORDER POLYMIXIIFORMES	BEARDFISHES
Polymixiidae	Beardfishes

ORDER PERCOPSIFORMES	TROUTPERCHES AND RELATIVES
Percopsidae	Troutperches
Aphredoderidae	Pirate perch
Amblyopsidae	Cavefishes

ORDER GADIFORMES	CODS, HAKES AND RELATIVES
Muraenolepididae	Muraenolepidids
Moridae	Morid cods (moras)
Melanonidae	Melanonids
Euclichthyidae	Euclichthyid
Bregmacerotidae	Codlets
Gadidae	Codfishes, haddocks and relatives
Merlucciidae	Hakes
Steindachneriidae	Steindachneriid
Macrouridae	Rattails (grenadiers)

ORDER OPHIDIIFORMES	CUSKEELS, PEARLFISHES AND RELATIVES
Ophidiidae	Cuskeels, brotulas and relatives
Carapidae	Pearlfishes
Bythitidae	Livebearing brotulas
Aphyonidae	Aphyonids

ORDER BATRACHOIDIFORMES	TOADFISHES AND MIDSHIPMEN
Batrachoididae	Toadfishes and midshipmen

ORDER LOPHIIFORMES	ANGLERFISHES, GOOSEFISHES AND FROGFISHES
Lophiidae	Goosefishes (monkfishes)
Antennariidae	Frogfishes (shallow anglerfishes)
Tetrabrachiidae	Humpback angler
Lophichthyidae	Lophichthyid
Brachionichthyidae	Warty anglers (hand fishes)
Chaunacidae	Seatoads (gapers, coffinfishes)
Ogcocephalidae	Batfishes
Caulophrynidae	Fanfin anglers
Ceratiidae	Seadevils
Gigantactinidae	Whipnose anglers
Neoceratiidae	Needlebeard angler
Linophrynidae	Netdevils
Oneirodidae	Dreamers
Thaumatichthyidae	Wolftrap angler
Centrophrynidae	Hollowchin anglers
Diceratiidae	Double anglers
Himantolophidae	Footballfishes
Melanocetidae	Blackdevils

ORDER GOBIESOCIFORMES

CLINGFISHES AND RELATIVES

Gobiesocidae — Clingfishes
Callionymidae — Dragonets
Draconettidae — Draconettids

ORDER CYPRINODONTIFORMES

KILLIFISHES AND RELATIVES

Rivulidae — South American annuals
Aplocheilidae — African annuals
Profundulidae — Profundulids
Fundulidae — Killifishes
Valenciidae — Valenciids
Anablepidae — Foureyed fishes (cuatro ojos)
Poeciliidae — Livebearers (guppies, etc.)
Goodeidae — Goodeids
Cyprinodontidae — Pupfishes

ORDER BELONIFORMES

RICEFISHES, FLYINGFISHES AND RELATIVES

Adrianichthyidae — Ricefishes
Exocoetidae — Flyingfishes
Hemiramphidae — Halfbeaks
Belonidae — Needlefishes
Scomberesocidae — Sauries

ORDER ATHERINIFORMES

SILVERSIDES, RAINBOWFISHES AND RELATIVES

Atherinidae — Silversides, topsmelts, grunions
Dentatherinidae — Dentatherinid
Notocheiridae — Surf silversides
Melanotaeniidae — Rainbowfishes
Pseudomugilidae — Blue eyes
Telmatherinidae — Sailfin silversides
Phallostethidae — Priapium fishes

ORDER STEPHANOBERYCIFORMES

PRICKLEFISHES AND RELATIVES

Stephanoberycidae — Pricklefishes
Melamphaidae — Bigscale fishes (ridgeheads)
Gibberichthyidae — Gibberfishes
Hispidoberycidae — Bristlyskin
Rondeletiidae — Orangemouth whalefishes
Barbourisiidae — Redvelvet whalefish
Cetomimidae — Flabby whalefishes
Mirapinnidae — Hairyfish, tapetails
Megalomycteridae — Mosaicscale (bignose) fishes

ORDER BERYCIFORMES

SQUIRRELFISHES AND RELATIVES

Holocentridae — Squirrelfishes (soldier fishes)
Berycidae — Alfoncinos
Anoplogasteridae — Fangtooth fishes
Monocentrididae — Pineapple (pinecone) fishes
Anomalopidae — Flashlight (lanterneye) fishes
Trachichthyidae — Roughies (slimeheads)
Diretmidae — Spinyfins

ORDER ZEIFORMES

DORIES AND RELATIVES

Parazenidae — Parazen
Macrurocyttidae — Macrurocyttids
Zeidae — Dories
Oreosomatidae — Oreos
Grammicolepididae — Tinselfishes
Caproidae — Boarfishes

ORDER GASTEROSTEIFORMES

STICKLEBACKS, PIPEFISHES AND RELATIVES

Hypoptychidae — Sand eel
Aulorhynchidae — Tubesnouts
Gasterosteidae — Sticklebacks
Indostomidae — Paradox fish
Pegasidae — Seamoths
Aulostomidae — Trumpetfishes
Fistulariidae — Cornetfishes
Macroramphosidae — Snipefishes
Centriscidae — Shrimpfishes (razorfishes)
Solenostomidae — Ghost pipefishes
Syngnathidae — Pipefishes, seahorses

ORDER SYNBRANCHIFORMES	SWAMPEELS AND RELATIVES
Synbranchidae	Swampeels
Mastacembelidae	Spiny eels
Chaudhuriidae	Chaudhuriids

ORDER SCORPAENIFORMES	SCORPIONFISHES AND RELATIVES
Scorpaenidae	Scorpionfishes (rockfishes), stonefishes

Caracanthidae	Coral crouchers
Aploactinidae	Velvetfishes
Pataecidae	Prowfishes
Gnathanacanthidae	Red velvetfish
Congiopodidae	Pigfishes (horsefishes)
Triglidae	Sea robins (gurnards)
Dactylopteridae	Helmet (flying) gurnards
Platycephalidae	Flatheads
Bembridae	Deepwater flatheads
Hoplichthyidae	Spiny (ghost) flatheads
Anoplopomatidae	Sablefish, skilfish
Hexagrammidae	Greenlings
Zaniolepididae	Combfishes
Normanichthyidae	Southern sculpin
Rhamphocottidae	Grunt sculpin
Ereuniidae	Deepwater sculpins
Cottidae	Sculpins
Comephoridae	Baikal oilfishes
Abyssocottidae	Abyssocottids
Hemitripteridae	Hemitripterids
Psychrolutidae	Flatheads (blob fishes)
Bathylutichthyidae	Bathylutichthyids
Agonidae	Poachers
Cyclopteridae	Lumpfishes
Liparidae	Snailfishes, lumpsuckers

ORDER PERCIFORMES	PERCHES AND RELATIVES
Suborder Percoidei	
Ambassidae	Glassfishes
Acropomatidae	Acropomatids
Epigonidae	Epigonids
Scrombropidae	Gnomefishes
Symphysanodontidae	Symphsanodontids
Caesioscorpididae	Caesioscorpidids
Polyprionidae	Wreckfishes
Dinopercidae	Cavebasses
Centropomidae	Snooks
Latidae	Giant perches
Percichthyidae	South temperate basses
Moronidae	North temperate basses
Serranidae	Sea basses, groupers
Callanthiidae	Rosey perches
Pseudochromidae	Dottybacks
Grammatidae	Basslets
Opistognathidae	Jawfishes
Plesiopidae	Roundheads
Notograptidae	Bearded snakeblennies
Pholidichthyidae	Convict blennies
Cepolidae	Bandfishes
Glaucosomatidae	Pearl perches
Terapontidae	Grunters
Banjosidae	Banjosids
Kuhliidae	Aholeholes
Centrarchidae	Sunfishes
Percidae	Perches, darters
Priacanthidae	Bigeyes (catalufas)
Apogonidae	Cardinalfishes
Dinolestidae	Dinolestid
Sillaginidae	Smelt whitings (whitings)
Malacanthidae	Tilefishes
Labracoglossidae	Labracoglossids
Lactariidae	False trevrelatives
Pomatomidae	Bluefish (tailor)
Menidae	Moonfish
Leiognathidae	Ponyfishes (slipmouths)
Bramidae	Pomfrets
Caristiidae	Manefishes
Arripidae	Australian salmons
Emmelichthyidae	Rovers
Lutjanidae	Snappers (emperors)
Lobotidae	Tripletails

Gerreidae	Mojarras (silver biddies)
Haemulidae	Grunts
Inermiidae	Bonnetmouths
Sparidae	Porgies (breams)
Centracanthidae	Centracanthids
Lethrinidae	Scavengers (emperors)
Nemipteridae	Threadfin, monocle breams
Sciaenidae	Drums, croakers
Mullidae	Goatfishes
Monodactylidae	Moonfishes (fingerfishes)
Pempherididae	Sweepers
Leptobramidae	Beachsalmon
Bathyclupeidae	Bathyclupeids
Toxotidae	Archerfishes
Coracinidae	Galjoenfishes
Kyphosidae	Seachubs
Girellidae	Nibblers (blackfishes)
Scorpididae	Halfmoons, mado (sweeps)
Microcanthidae	Stripey
Ephippidae	Spadefishes
Scatophagidae	Scats
Chaetodontidae	Butterflyfishes
Pomacanthidae	Angelfishes
Enoplosidae	Oldwife
Pentacerotidae	Armorheads
Oplegnathidae	Knifejaws
Icosteidae	Ragfish
Kurtidae	Nurseryfishes
Scombrolabracidae	Scombrolabracid

Suborder Carangoidei

Nematistiidae	Roosterfish
Carangidae	Jacks (trevallies)
Echeneidae	Remoras
Rachycentridae	Cobia (black kingfish)
Coryphaenidae	Dolphins

Suborder Cirrhitoidei

Cirrhitidae	Hawkfishes
Chironemidae	Kelpfishes
Aplodactylidae	Aplodactylids
Cheilodactylidae	Morwongs
Latrididae	Trumpeters

Suborder Mugiloidei

Mugilidae	Mullets

Suborder Polynemoidei

Polynemidae	Threadfins

Suborder Labroidei

Chichlidae	Cichlids
Embiotocidae	Surfperches
Pomacentridae	Damselfishes, anemonefishes
Labridae	Wrasses
Odacidae	Rock whitings
Scaridae	Parrotfishes

Suborder Zoarcoidei

Bathymasteridae	Ronquils
Zoarcidae	Eelpouts
Stichaeidae	Pricklebacks
Cryptacanthodidae	Wrymouths
Pholididae	Gunnels
Anarhichadidae	Wolffishes
Ptilichthyidae	Quillfishes
Zaproridae	Prowfish
Scytalinidae	Graveldiver

Suborder Notothenioidei

Bovichtidae	Thornfishes
Nototheniidae	Nototheniids
Harpagiferidae	Plunderfishes
Bathydraconidae	Antarctic dragonfishes
Channichthyidae	Icefishes

Suborder Trachinoidei

Chiasmodontidae	Swallowers
Champsodontidae	Champsodontids
Trichodontidae	Sandfishes
Trachinidae	Weeverfishes
Uranoscopidae	Stargazers
Trichonotidae	Sanddivers
Creediidae	Sandburrowers
Leptoscopidae	Leptoscopids
Percophidae	Duckbills
Pinguipedidae	Sandperches
Cheimarrhichthyidae	Torrentfish
Ammodytidae	Sandlances

Suborder Blennioidei

Tripterygiidae — Triplefins
Dactyloscopidae — Sand stargazers
Labrisomidae — Labrisomids
Clinidae — Kelp blennies (kelpfishes)
Chaenopsidae — Tube blennies
Blenniidae — Combtooth blennies

Suborder Gobiodei

Rhyacichthyidae — Loach goby
Odontobutidae — Freshwater Asian gobies
Gobiidae — Gobies, sleepers (gudgeons)

Suborder Acanthuroidei

Siganidae — Rabbitfishes
Luvaridae — Louvar
Zanclidae — Moorish idol
Acanthuridae — Surgeonfishes (tangs)

Suborder Scombroidei

Sphyraenidae — Barracudas
Gempylidae — Snake mackerels, gemfishes
Trichiuridae — Cutlassfishes
Scombridae — Mackerels, tunas, bonitos
Xiphiidae — Swordfish
Istiophoridae — Billfishes

Suborder Stromateoidei

Amarsipidae — Amarsipids
Centrolophidae — Medusafishes
Nomeidae — Driftfishes
Ariommatidae — Ariommatids
Tetragonuridae — Squaretails
Stromateidae — Butterfishes

Suborder Anabantoidei

Badidae — Chameleonfish
Nandidae — Leaffishes
Pristolepidae — False leaffishes
Channidae — Snakeheads
Anabantidae — Climbing gouramies
Belontiidae — Gouramies
Helostomatidae — Kissing gouramie
Osphronemidae — Giant gouramie
Luciocephalidae — Pikehead

ORDER PLEURONECTIFORMES — FLATFISHES

Psettodidae — Spiny flatfishes
Citharidae — Citharids
Bothidae — Lefteye flounders
Pleuronectidae — Tonguefishes (tongue soles)
Soleidae — Soles
Achiridae — American soles

ORDER TETRAODONTIFORMES — TRIGGERFISHES AND RELATIVES

Triacanthodidae — Spikefishes
Triacanthidae — Triplespines
Balistidae — Triggerfishes
Monacanthidae — Filefishes (leatherjackets)
Aracanidae — Robust boxfishes
Ostraciidae — Boxfishes, cowfishes, trunkfishes
Triodontidae — Pursefish (three-toothed puffer)
Tetraodontidae — Pufferfishes (toados)
Diodontidae — Porcupinefishes
Molidae — Ocean sunfishes (molas)

Invertebrates

PHYLUM PORIFERA — SPONGES
- Class Calcarea — Calcareous sponges
- Class Hexactinellida — Glass sponges
- Class Demospongiae — Demosponges

PHYLUM CHORDATA — INVERTEBRATE CHORDATES
- Class Urochordata — Sea squirts or tunicates
- Class Cephalochardata — Lancelets

PHYLUM CNIDARIA — HYDRAS, JELLYFISH, SEA ANEMONES, CORALS
- Class Hydrozoa — Hydras and freshwater jellyfish
- Class Scyphozoa — Jellyfish
- Class Cubozoa — Jellyfish
- Class Anthozoa — Sea anemones, corals, sea fans, sea pansies

PHYLUM CTENOPHORA — SEA WALNUTS OR COMB JELLIES
- Class Tentaculata
- Class Nuda

PHYLUM PLATYHELMINTHES — FLATWORMS
- Class Trematoda — Flukes
- Class Monogenea
- Class Cestoda — Tapeworms
- Class Turbellaria — Free-living flatworms
- Subclass Archoophora — Archoophoran turbellarians
- Subclass Neoophora — Neoophoran turbellarians

PHYLUM GNATHOSTOMULIDA — ACOELOMATE WORMS
- Phylum Gnathostomulida — Acoelomate worms

PHYLUM ORTHONECTIDA
- Phylum Orthonectida

PHYLUM RHOMBOZOA
- Class Dicyemida
- Class Heterocyemida

PHYLUM NEMERTEA — RIBBON OR PROBOSCIS WORMS
- Class Anopla — Unarmed nemerteans
- Class Enopla — Armed nemerteans

PHYLUM GASTROTRICHA — MARINE AND FRESHWATER GASTROTRICHS
- Phylum Gastrotricha — Marine and freshwater gastrotrichs

PHYLUM NEMATODA — ROUNDWORMS
- Class Adenophorea
- Class Secernentea

PHYLUM NEMATOMORPHA — HORSEHAIR OR HAIRWORMS
- Class Nectonematoida — Marine nematomorphs
- Class Gordioida — Freshwater and semiterrestrial nematomorphs

PHYLUM ROTIFERA — ROTIFERS
- Class Seisonidea
- Class Bdelloidea
- Class Monogononta

PHYLUM ACANTHOCEPHALA — ENDOPARASITES
- Class Archiacanthocephala — Parasites of birds and mammals
- Class Eoacanthocephala — Parasites of fishes and reptiles
- Class Palaeacanthocephala — Parasites of all vertebrates

PHYLUM KINORHYNCHA
- Order (class) Cyclorhagida
- Order (class) Homalorhagida

PHYLUM LORICIFERA
- Phylum Loricifera

PHYLUM TARDIGRADA — WATER BEARS
Class Heterotardigrada
Class Eutardigrada

PHYLUM SIPUNCULA — PEANUT WORMS
Phylum Sipuncula — Peanut worms

PHYLUM ECHIURA — SPOONWORMS
Phylum Echiura — Spoonworms

PHYLUM PRIAPULIDA
Phylum Priapulida

PHYLUM MOLLUSCA — MOLLUSKS
Class Aplacophora — Wormlike mollusks
Class Polyplacophora — Chitons
Class Monoplacophora
Class Gastropoda — Gastropods
Class Bivalvia — Bivalves
Class Scaphopoda
Class Cephalopoda — Squid and octopi

PHYLUM ANNELIDA — SEGMENTED WORMS
Class Polychaeta
Class Oligochaeta
Class Hrudinea — Leeches
Class Branchiobdellida

PHYLUM POGONOPHORANS
Phylum Pogonophorans

PHYLUM ARTHROPODA — ARTHROPODS
Class Insecta — Insects
  Subclass Entognatha — Wingless insects
  Subclass Ectognatha — Winged insects

SUBPHYLUM CHELICERATA
Class Merostomata
Class Arachnida — Spiders, scorpions, mites and ticks
Class Pycnogonida — Sea spiders

SUBPHYLUM CRUSTACEA — CRUSTACEANS
Class Remipedia
Class Cephalocarida
Class Branchiopoda
Class Ostracoda — Ostracods
Class Copepoda
Class Mystacocarida
Class Tantulocarida
Class Branchiura
Class Cirripedia — Barnacles
Class Pentastomida
Class Malacostraca — Crabs, lobsters, shrimp

SUBPHYLUM MYRIAPODA — MYRIAPODUS ARTHROPODS
Class Chilopoda — Centipedes
Class Symphyla
Class Diplopoda — Millipedes
Class Pauropoda — Pauropods

PHYLUM ONYCHOPHORA
Phylum Onychophora

PHYLUM HEMICHORDATA
Class Enteropneusta — Acorn worms
Class Pterobranchia

PHYLUM CHORDATA

SUBPHYLUM UROCHORDATA
Class Ascidiacea — Sea squirts
Class Thaliacea
Class Larvacea

SUBPHYLUM CEPHALOCHORDATA	LANCELETS

PHYLUM CHAETOGNATHA ARROWWORMS

Phylum Chaetognatha	Arrowworms

PHYLUM ECHINODERMATA	ECHINODERMS

SUBPHYLUM HOMALOZOA

SUBPHYLUM CRINOZOA

Class Crinoidea	Sea lilies, feather stars

SUBPHYLUM ASTEROZOA

Class Asteroidea	Sea stars
Class Ophiuroidea	Brittle stars
Class Concentricycloidea	

SUBPHYLUM ECHINOZOA

Class Echinoidea	Sea urchins, sand dollars
Class Holothuroidea	Sea cucumbers

PHYLUM BRYOZOA

Class Phylactolaemata
Class Stenolaemata
Class Gymnolaemata

PHYLUM ENTOPROCTA

Phylum Entoprocta

PHYLUM PHORONIDA

Phylum Phoronida

PHYLUM BRACHIOPODA	BRACHIPODS

Class Inarticulata
Class Articulata

CLASS ARACHNIDA

ORDER ARANEAE	SPIDERS
Agelenidae	Funnel weavers
Antrodiaetidae	Folding trapdoor spiders
Araneidae	Orb weavers
Clubionidae	Sac spiders
Ctenidae	Wandering spiders
Ctenizidae	Trapdoor spiders
Dictynidae	Dictynid spiders
Dipluridae	Funnel-web spiders
Eresidae	Eresid spiders
Heteropodidae	Huntsman spiders
Linyphiidae	Dwarf spiders
Loxoscelidae	Violin spiders
Lycosidae	Wolf spiders
Oxyopidae	Lynx spiders
Pholcidae	Daddy-long-legs spiders
Pisauridae	Nursery-web spiders
Salticidae	Jumping spiders
Scytodidae	Spitting spiders
Selenopidae	Selenopid crab spiders
Sparassidae	Giant crab spiders
Tetragnathidae	Large-jawed orb weavers
Theraphosidae	Tarantulas
Theridiidae	Comb-footed spiders
Thomisidae	Crab spiders

ORDER SCORPIONES	SCORPIONS
Buthidae	Buthid scorpions
Iuridae	Iurid scorpions

ORDER PSEUDOSCORPIONES	PSEUDO-SCORPIONS
Cheliferidae	
Chernetidae	Chernetids

ORDER OPILIONES	HARVESTMEN
Leiobunidae	
Phalangiidae	

ORDER ACARINA	MITES AND TICKS
Argasidae	Soft ticks
Hydrachnellae	Water mites
Ixodidae	Hard ticks
Tetranychidae	Spider mites
Trombidiidae	Velvet mites

ORDER UROPYGI	WHIPSCORPIONS
Thelyphonidae	Vinegaroons

ORDER AMBLYPYGI	TAILLESS WHIPSCORPIONS
Phrynidae	
Tarantulidae	

ORDER SOLIFUGAE	WINDSCORPIONS
Eremobatidae	
Solpugidae	

CLASS INSECTA

ORDER PROTURA	PROTURANS
Eosentomidae	

ORDER COLLEMBOLA	SPRINGTAILS
Entomobryidae	
Hypogastruridae	
Isotomidae	
Onychiuridae	
Sminthuridae	Globular springtails

ORDER DIPLURA	DIPLURANS
Campodeidae	Campodeids
Japygidae	Japygids

ORDER ARCHAEOGNATHA	BRISTLETAILS
Machilidae	Jumping bristletails
Meinertellidae	

ORDER THYSANURA	SILVERFISH
Lepismatidae	Silverfish (firebrats)

ORDER EPHEMEROPTERA	MAYFLIES
Baetidae	Small mayflies
Ephemerellidae	Midboreal mayflies
Ephemeridae	Burrowing mayflies
Heptageniidae	Stream mayflies
Leptophlebiidae	Spinners

ORDER ODONATA	DRAGONFLIES AND DAMSELFLIES
Aeschnidae	Darners
Calopterygidae	Broad-winged damselflies
Coenagrionidae	Narrow-winged damselflies
Cordulegastridae	Biddies
Corduliidae	Green-eyed skimmers
Gomphidae	Clubtails
Lestidae	Spread-winged damselflies
Libellulidae	Common skimmers
Macromiidae	Belted and river skimmers
Petaluridae	Graybacks

ORDER BLATTODEA	COCKROACHES
Blaberidae	
Blattellidae	
Blattidae	Common cockroaches

ORDER MANTODEA	MANTIDS
Empusidae	
Hymenopodidae	Flower mantids
Mantidae	Common praying mantids

ORDER ISOPTERA	TERMITES
Hodotermitidae	Rotten-wood termites
Kalotermitidae	Damp-wood termites
Mastotermitidae	
Rhinotermitidae	Subterranean termites
Termitidae	Nasutiform termites

ORDER ZORAPTERA — ZORAPTERANS
Zorotypidae — Zorapterans

ORDER GRYLLOBLATTODEA — ICE INSECTS
Grylloblattidae — Rock crawlers

ORDER DERMAPTERA — EARWIGS
Chelisochidae — Black earwigs
Forficulidae — Common earwigs
Labiduridae — Long-horned earwigs
Labiidae — Little earwigs

ORDER PLECOPTERA — STONEFLIES
Capniidae — Small winter stoneflies
Chloroperlidae — Green stoneflies
Isoperlidae — Green-winged stoneflies
Leuctridae — Rolled-winged stoneflies
Nemouridae — Spring stoneflies
Peltoperlidae — Roachlike stoneflies
Perlidae — Common stoneflies
Perlodidae — Perlodid stoneflies
Pteronarcidae — Giant stoneflies
Taeniopterygidae — Winter stoneflies

ORDER ORTHOPTERA — GRASSHOPPERS AND CRICKETS
Acrididae — Short-horned grasshoppers
Cooloolidae — Cooloola monster
Cylindrachetidae — Sand gropers
Eneopterinae — Bush crickets
Eumastacidae; Tanaoceridae — Monkey grasshoppers
Gryllacrididae; Rhaphidophoridae — Leaf-rolling crickets, camel crickets, cave crickets
Gryllidae — True crickets
Gryllotalpidae — Mole crickets
Myrmecophilidae — Ant crickets
Oecanthinae Pyrgomorphidae — Tree crickets
Stenopelmatidae — King crickets
Tetrigidae — Pygmy grasshoppers
Tettigoniidae — Long-horned grasshoppers and katydids
Tridactylidae — Pygmy mole crickets

ORDER PHASMATODEA — STICK AND LEAF INSECTS
Phasmatidae — Stick insects
Phasmidae — Walking sticks
Timemidae — Timemas

ORDER EMBIOPTERA — WEBSPINNERS
Clothodidae — Clothodids
Embiidae — Embiids

ORDER PSOCOPTERA — BOOKLICE AND BARKLICE
Lepidopsocidae — Scaly barklice
Liposcelididae — Liposcelid booklice
Pseudocaeciliidae
Psocidae — Common barklice
Psyllipsocidae — Psyllipsocids
Trogiidae — Trogiid booklice

ORDER PHTHIRAPTERA — PARASITIC LICE
Boopidae
Echinophthiriidae — Spiny sucking lice
Gyropidae — Guinea pig lice
Haematopinidae — Mammal-sucking lice
Hoplopleuridae
Laemobothriidae — Bird lice
Linognathidae — Smooth sucking lice
Menoponidae — Poultry-chewing lice
Pediculidae — Human lice
Philopteridae — Feather-chewing lice
Ricinidae — Bird lice
Trichodectidae — Mammal-chewing lice

ORDER HEMIPTERA — BUGS
Achilidae — Achilid planthoppers
Adelgidae — Pine aphids
Aleyrodidae — Whiteflies
Alydidae — Broad-headed bugs
Anthocoridae — Flower bugs, minute pirate bugs
Aphididae Aphrophoridae — Aphids
Aradidae — Flat bugs, bark bugs
Asterolecaniidae — Pit scales
Belostomatidae — Giant water bugs

Berytidae	Stilt bugs
Carsidaridae	
Cercopidae	Spittlebugs and froghoppers
Chermidae	Pine and spruce aphids
Cicadellidae	Leafhoppers
Cicadidae	Cicadas
Cimicidae	Bed bugs
Cixiidae	Cixiid planthoppers
Coccidae; Colobathristidae	Soft scales, wax scales
Coreidae	Leaf-footed bugs, crusader bugs
Corixidae	Water boatmen
Cydnidae	Negro bugs
Dactylopiidae	Cochineal bugs
Delphacidae	Delphacid planthoppers
Derbidae	Derbid planthoppers
Diaspididae	Armored scale insects
Dictyopharidae; Dinidoridae	Dictyopharid planthoppers
Dipsocoridae; Schizopteridae	Jumping ground bugs
Eriosomatidae	Woolly and gall-making aphids
Eurymelidae; Membracidae	Treehoppers
Flatidae	Flatid planthoppers
Fulgoridae	Fulgorids
Gelastocoridae	Toad bugs
Gerridae	Water striders
Hebridae Homotomidae	Velvet water bugs
Hydrometridae	Water measurers
Isometopidae	Jumping tree bugs
Issidae	Issid planthoppers
Kermidae	Gall-like coccids
Kerriidae; Lacciferidae	Lac insects
Leptopodidae	Spiny shore bugs
Lygaeidae	Seed bugs
Margarodidae	Giant scale insects
Mesoveliidae	Water treaders
Miridae	Leaf or plant bugs
Nabidae	Damsel bugs
Naucoridae	Creeping water bugs
Nepidae	Waterscorpions
Notonectidae	Backswimmers
Ochteridae	Velvety shore bugs
Ortheziidae	Ensign scales
Peloridiidae	Moss bugs
Pentatomidae	Stink bugs, shield bugs
Phylloxeridae	Gall aphids
Phymatidae	Ambush bugs
Polyctenidae	Bat bugs
Pseudococcidae; Eriococcidae	Mealybugs
Psyllidae	Psyllids, lerps
Pyrrhocoridae	Red bugs, stainers
Reduviidae	Assassin bugs
Rhopalidae	Scentless plant bugs
Ricaniidae	Ricaniid planthoppers
Saldidae	Shore bugs
Scutelleridae	Jewel bugs, shield-backed bugs
Tessaratomidae	
Tettigarctidae	Hairy cicadas
Tingidae	Lace bugs
Triozidae	
Veliidae	Ripple bugs

ORDER THYSANOPTERA — THRIPS

Aeolothripidae	Banded thrips
Phloeothripidae	Tube-tailed thrips
Thripidae	Common thrips

ORDER MEGALOPTERA — ALDERFLIES AND DOBSONFLIES

Corydalidae	Dobsonflies
Sialidae	Alderflies

ORDER RAPHIDIOPTERA — SNAKEFLIES

Inocelliidae	
Raphidiidae	

ORDER NEUROPTERA — NET-VEINED INSECTS

Ascalaphidae	Owlflies
Chrysopidae	Green lacewings
Coniopterygidae	Dusty-wings
Hemerobiidae	Brown lacewings
Ithonidae	Moth lacewings
Mantispidae	Mantidflies
Myrmeleontidae	Antlions
Nemopteridae	
Polystoechotidae	Giant lacewings
Psychopsidae	Silky lacewings
Sisyridae	Spongeflies

ORDER COLEOPTERA	BEETLES
Anobiidae	Furniture beetles
Anthribidae	Fungus weevils
Bostrichidae	Branch and twig borers, auger beetles
Brentidae	Primitive weevils
Bruchidae	Seed beetles
Buprestidae	Metallic wood-boring beetles, jewel beetles
Cantharidae	Soldier beetles
Carabidae	Ground beetles
Cerambycidae	Longhorn beetles, longicorn beetles
Chrysomelidae	Leaf beetles
Cicindelidae	Tiger beetles
Cleridae	Checkered beetles
Coccinellidae	Ladybird beetles
Cucujidae	Flat bark beetles
Curculionidae	Snout beetles and weevils
Dermestidae	Dermestid beetles
Dytiscidae	Predacious diving beetles
Elateridae	Click beetles
Erotylidae	Pleasing fungus beetles
Gyrinidae	Whirligig beetles
Haliplidae	Crawling water beetles
Histeridae	Hister beetles
Hydrophilidae; Laemophloeidae	Water scavenger beetles
Lampyridae	Fireflies, lightning bugs
Lathridiidae	Minute brown scavenger beetles
Lucanidae	Stag beetles
Lycidae	Net-winged beetles
Lymexylidae	Ship-timber beetles
Meloidae	Blister beetles
Melyridae	Softwinged flower beetles
Mycetophagidae	Hairy fungus beetles
Nitidulidae	Nitidulid beetles
Passalidae	Passalid beetles or bessbugs
Psephenidae	Water pennies
Ptiliidae	Feather-winged beetles
Ptinidae	Spider beetles
Pyrochroidae	Fire-colored beetles
Rhipiphoridae	Rhipiphoridan beetles
Scarabaeidae	Scarab beetles
Scolytidae	Bark and ambrosia beetles
Silphidae	Carrion beetles
Silvanidae	Flat grain beetles
Staphylinidae	Rove beetles
Tenebrionidae	Darkling beetles
Trogidae	Carcass beetles
Trogositidae	Bark-gnawing beetles

ORDER STREPSIPTERA	TWISTED-WINGED PARASITES
Mengeidae	
Stylopidae	Stylopids

ORDER MECOPTERA AND RELATIVES	SCORPIONFLIES
Bittacidae	Hangingflies
Boreidae	Snow scorpionflies
Panorpidae	Common scorpionflies

ORDER SIPHONAPTERA	FLEAS
Dolichopsyllidae; Ceratophyllidae	Rodent fleas
Leptopsyllidae	Mouse fleas
Pulicidae	Common fleas
Tungidae	Sticktight and chigoe fleas

ORDER DIPTERA	FLIES
Acroceridae	Bladder flies
Agromyzidae	Leafminer flies
Anthomyiidae	Anthomyiid flies
Apioceridae	Flower-loving flies
Asilidae	Robber flies
Bibionidae	March flies
Blephariceridae	Net-winged midges
Bombyliidae	Bee flies
Braulidae	Beelice
Calliphoridae	Blowflies
Canacidae	Beach flies
Cecidomyiidae	Gall midges
Ceratopogonidae	Punkies, biting midges
Chaoboridae	Phantom midges
Chironomidae	Midges
Chloropidae	Grass flies

Coelopidae	Seaweed flies
Conopidae	Thick-headed flies
Culicidae	Mosquitoes
Dolichopodidae	Long-legged flies
Drosophilidae	Vinegar (pomace) flies
Empididae	Dance flies, water cruisers
Ephydridae	Shore flies
Fanniidae	
Fergusoninidae	Eucalyptus flies
Heleomyzidae	Sun flies
Hippoboscidae	Louse flies and sheep keds
Lonchaeidae	Lance flies
Micropezidae	Stilt-legged flies
Muscidae	House flies and bushflies
Mycetophilidae	Fungus gnats
Mydidae	Mydas flies
Nemestrinidae	Tanglevein flies
Nerriidae	Cactus flies
Neurochaetidae	Upside-down flies
Nycteribiidae; Streblidae	Bat flies
Oestridae; Gasterophilidae	Botflies
Phoridae	Scuttle flies
Piophilidae	Skipper flies
Platypezidae	Flat-footed flies
Platystomatidae	Platystomatid flies
Psychodidae	Moth flies
Ptychopteridae	Phantom crane flies
Pyrgotidae	Pyrgotid flies
Rhagionidae	Snipe flies
Sarcophagidae	Flesh flies
Scenopinidae	Window flies
Sciaridae	Black fungus gnats
Sciomyzidae	March flies
Sepsidae	Ant flies
Simuliidae	Black or sand flies
Sphaeroceridae	Short heel flies, small dung flies
Stratiomyidae	Soldier flies
Syrphidae	Hover flies
Tabanidae	Horse and deer flies
Tachinidae	Tachinid flies
Tephritidae	Fruit flies
Teratomyzidae	Fern flies
Therevidae	Stiletto flies
Tipulidae	Crane flies
Trichoceridae	Winter crane flies

ORDER TRICHOPTERA — CADDISFLIES

Hydropsychidae	Net-spinning caddisflies
Leptoceridae	Long-horned caddisflies
Limnephilidae	Northern caddisflies
Phryganeidae	Large caddisflies
Psychomyiidae	Tube-making caddisflies

ORDER LEPIDOPTERA — BUTTERFLIES AND MOTHS

Alucitidae	Many-plume moths
Anthelidae	
Apaturidae	Hackberry and goatweed butterflies
Arctiidae	Tiger moths
Bombycidae	Silkworm moths
Bucculatricidae	
Carposinidae	Carposinid moths
Carthaeidae	
Castniidae	
Citheroniidae	Royal moths
Coleophoridae	Casebearer moths
Cosmopterygidae	
Copromorphidae	
Cossidae	Wood moths
Ctenuchidae	Ctenuchid moths
Danaidae	Milkweed butterflies
Drepanidae	Hook-tip moths
Eupterotidae; Zanolidae	Zanolid moths
Gelechiidae	Gelechiid moths
Geometridae	Measuringworm moths, loopers
Gracilariidae	Leaf blotch miners
Hepialidae	Ghost moths, swifts
Hesperiidae	Skippers
Incurvariidae	Yucca moths and relatives
Lasiocampidae	Tent caterpillar and lappet moths
Libytheidae	Snout butterflies
Limacodidae	
Lycaenidae	Gossamer-winged butterflies
Lymantriidae	Tussock moths and relatives
Lyonetiidae	Lyonetiid moths
Noctuidae	Noctuids
Notodontidae	Prominents
Nymphalidae	Brush-foots

Oecophoridae	Oecophorid moths
Papilionidae	Swallowtails and relatives
Pieridae	Whites and sulfurs
Plutellidae	Diamondback moths
Pterophoridae	Plume moths
Psychidae	Bagworm moths
Pyralidae	Pyralid moths
Riodinidae	Metalmarks
Saturniidae	Giant silkworm moths
Satyridae	Satyrs, nymphs, and arctics
Sesiidae	Clear-winged moths
Sphingidae	Sphinx or hawk moths
Thaumetopoeidae	
Tineidae	Clothes moths and relatives
Tortricidae	Tortricid moths
Uraniidae	
Yponomeutidae	Ermine moths
Zygaenidae	Smoky moths

ORDER HYMENOPTERA — BEES, ANTS, WASPS AND SAWFLIES

Agaonidae	Fig wasps
Andrenidae	Andrenid bees
Apidae	Digger bees, carpenter bees, cuckoo bees, bumblebees, honeybees and relatives
Argidae	Argid sawflies
Braconidae	Braconids
Cephidae	Stem sawflies
Chalcididae	Chalcids
Chrysididae	Cuckoo wasps
Cimbicidae	Cimbicid sawflies
Colletidae	Yellow-faced and plasterer bees
Cynipidae	Gall wasps
Encyrtidae	Encyrtid wasps
Eulophidae; Aphelininae	Eulophid wasps
Eurytomidae	Seed chalcids
Evaniidae	Hatchet wasps
Formicidae	Ants
Halictidae	Sweat (halictid) bees
Ibaliidae	Ibaliid wasps
Ichneumonidae	Ichneumonids
Megachilidae	Leaf-cutting bees
Melittidae	Melittid bees
Mutillidae	Velvet ants
Mymaridae	Fairy wasps
Pelecinidae	Pelecinids
Pergidae	Pergid sawflies
Pompilidae	Spider wasps
Pteromalidae	Pteromalid wasps
Scelionidae	Scelionid wasps
Scoliidae	Scoliid wasps
Siricidae	Horntails
Sphecidae	Sphecid wasps
Stephanidae	Stephanid wasps
Symphyta	Sawflies
Tenthredinidae	Common sawflies
Tiphiidae	Tiphid wasps
Torymidae	Torymid wasps
Vespidae	Vespid wasps

Size Comparisons

COMPARING MAMMALS

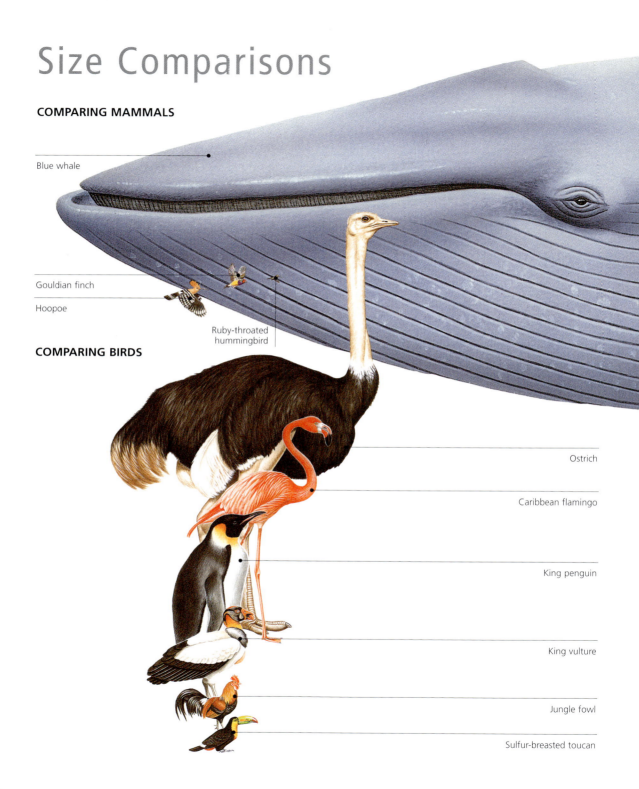

Blue whale

Gouldian finch

Hoopoe

Ruby-throated hummingbird

COMPARING BIRDS

Ostrich

Caribbean flamingo

King penguin

King vulture

Jungle fowl

Sulfur-breasted toucan

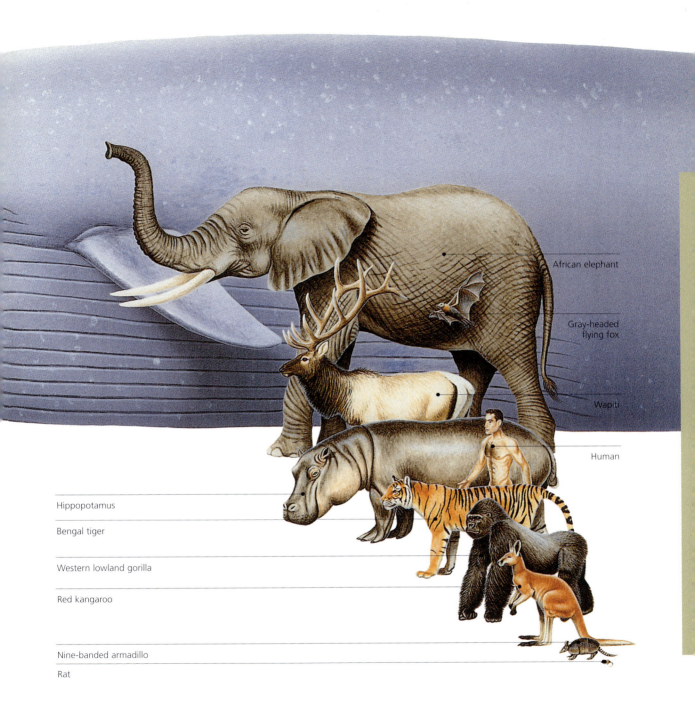

Size Comparisons

COMPARING REPTILES

Indopacific crocodile

Green anaconda

COMPARING FISHES

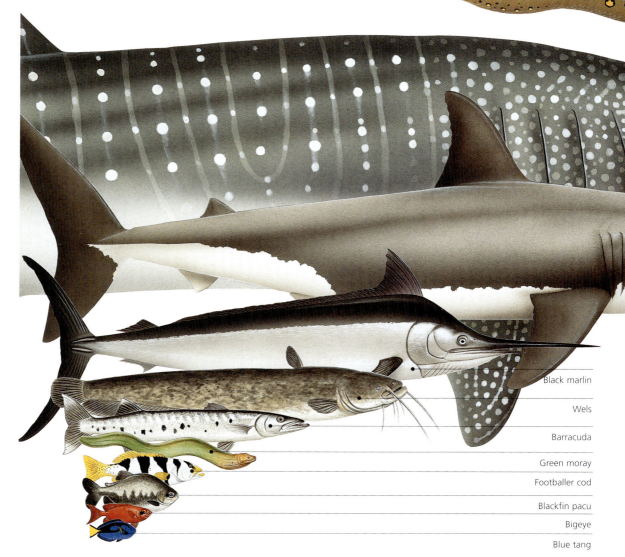

Black marlin

Wels

Barracuda

Green moray

Footballer cod

Blackfin pacu

Bigeye

Blue tang

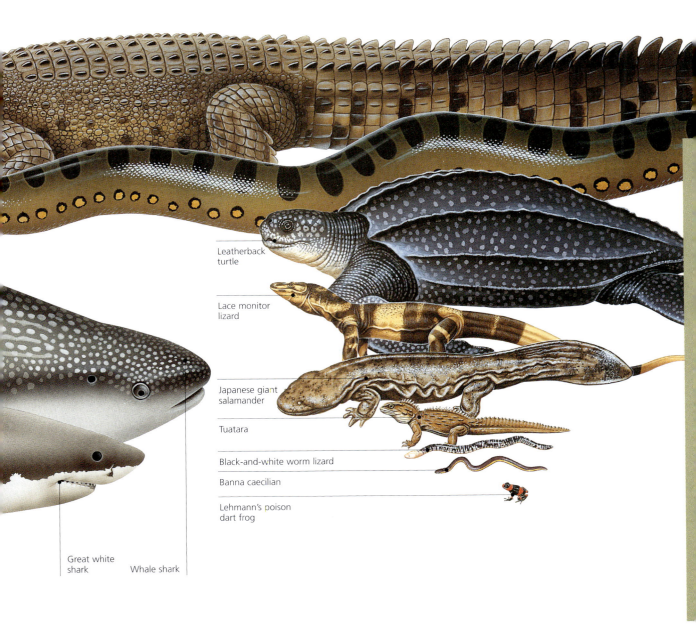

Index

A

aardvark 79, 206
aardwolf 145
Ablabys taenionotus 479
Acanthaster planci 501
Acanthophis antarcticus 395
Acanthorhynchus tenuirostris 326
Acantophthalmus myersi 460
Accrididae family 524
Acinonyx jubatus 155
Acipenser
    oxyrinchus 452
    ruthenus 453
Aconaemys fuscus 227
Acrochordus
    arafurae 386
    granulatus 387
Acryllium vulturinum 277
Actitis macularia 284
adder, common death 395
Aegotheles sp. 299
Agalychnis callidryas 417
agama 366
    Iranian toad-headed 366
Agama agama 366
agouti, Brazilian 226
Ahaetulla nasuta 386
Ailuroedus dentirostris 335
Ailuropoda melanoleuca 130, 137
Ailurus fulgens 145
Aix sponsa 262
Ajaia ajaia 256
Alauda arvensis 318
albatross
    gray-headed 250
    wandering 237
Albertosaurus 57
Alcedo cristata 306
Alces alces 191
Allenopithecus nigroviridis 112

Alligator
    mississippiensis 356
    sinesis 357
alligators 343, 356-7
Allosaurus 43, 47, 51, 57, 59
Alopex lagopus 128
Alopias vulpinus 441
Alosa sapidissima 459
Alouatta seniculus 116
alpaca 184-5
Alphadon 66
alpine regions 26-7
Amblyornis macgregoriae 335
Amblyrhynchus cristatus 365
Ambystoma
    laterale 404
    mexicanum 406
    tigrinum 405
Ameiva ameiva 379
Amia calva 452
ammonite 39
amphibians
    classifying 400-1, 563
amphisbaenians 360-1
Amphiuma means 406
Anableps anableps 475
anaconda 582-3
Anas
    cyanoptera 263
    rubripes 263
Anatotitan 60
anchoveta, Peruvian 458
Andrias
    alpestris 407
    japonicus 407
anemonefish, spinecheek 483
angler, Paxton's whipnose 471
Anguilla rostrata 456
Anilius scytale 383
anole
    Cuban brown 365
    dewlap 362
Anolis sagrei sagrei 365

Anopheles sp. 538
Anostomus anostomus 463
Anser anser 262
anteater, giant 93
antelopes 193-5
    four-horned 195
    sable 194
    saiga 195
antennae 519
Anthus spinoletta 318
Antilocapra americana 187
Antilope cervicapra 195
antlers 188-9
ants 546-9
    Amazon 549
    bulldog 546
    leaf-cutter 549
    red 549
Aotus trivirgatus 116
apalis, bar-throated 314
Apaloderma narina 304
Apalone spinifera spinifera 353
apes 118-21 *see also* monkeys, primates
Aphelocoma coerulescens 337
Aphredoderus sayanus 470
Aphyosemion gardneri 475
Apidae family 546
Aplocheilus annulatus 475
Apus affinis 302
aracari, curl-crested 313
Aramus guarauna 278
arawana 454
Archaeopteryx 43, 66, 72-3, 240
Archelon 67, 347
Archilochus
    alexandri 302
    colubris 302
Architeuthis sp. 496
Arctiidae family 541
Arctocebus calabarensis 108
Arctonyx collaris 139

Ardea
  cinerea 256
  goliath 257
  purpurea 257
armadillo 78, 92, 581
Arsinoitherium 83
arthropods 502–13
Arthur's paragalaxias 464
Ascaphus truei 412
ass 205
Astronotus ocellatus 483
Atelopus varius 416
Athene noctua 297
Atta sp. 549
Aulacorhynchus prasinus 313
Aulostomus maculatus 478
Australophocaena dioptrica 177
Aviceda subcristata 269
avocet 282
axolotl, Mexican 401, 406
aye-aye 81, 109

B

babblers 323
babirusa 181
baboon 104
  gelada 18, 112
  hamadryas 113
Babyrousa babyrussa 181
badgers 138–9
Balaena mysticetus 169
Balaenoptera
  acutorostrata 168
  musculus 169
  physalus 168
Balantiocheilos melanopterus 461
Balistoides conspicillum 485
barasingha 190

barbet
  double-toothed 313
  red-and-yellow 312
barramundi 426
Baryonyx 57
Basiliscus
  basiliscus 369
  plumifrons 368
basilisk, green 368
Bassariscus astutus 145
bateleur 269
Batrachostomus javensis 298
bats 98–9
  African yellow-winged 102
  Egyptian fruit 101
  Gambian epauletted fruit 100
  greater bulldog 103
  hammer-headed fruit 100
  Honduran white 103
  horseshoe 102
  insect-eating 102–3
  Old World fruit 100–1
  spotted 102
  sword-nosed 103
  tent-building 103
  vampire 78, 102
  wrinkle-faced 103
baza, Pacific 269
bears 130–1
  American black 130–3
  Asiatic black 130, 136
  brown, 122, 123, 130–1
  Eurasian and New World 132–3
  polar 130, 134–5
  sloth 130, 136
  spectacled 130, 133
  sun 130, 137
beavers 79, 216, 222–3
  Eurasian 223
  mountain 216, 217
Bedotia geayi 475
bee-eaters 307, 308, 314

bees 518, 546–7
  bumblebee 546
  honey 518, 550–1
beetles 518, 530–1
  atlas 490
  bombardier 533
  Colorado 531
  darkling 530
  diving 530, 533
  fire 530
  flea 531
  harlequin 532
  Hercules 530–1
  jewel 530
  long-horned 532
  rhinoceros 531
  rove 533
  tortoise 532
  violin 533
  whirligig 530
beluga 453
bichirs 452–3
  mottled 427
bilby 89
bills 283, 290
bird-of-paradise 314, 334
birds *see also by name*
  anatomy 242–3
  characteristics 238–9
  classifying 236–7, 558–61
  evolution 240–1
bison 198, 342
Bison
  bison 198
  bonasus 198
Bitis nasicornis 395
bitterling 460
bittern, great 254
blackbird, Eurasian 314
blackbuck 195
blackfish, Alaska 467
blenny, black-headed 482
bluebird 323

boa, Brazilian rainbow 385
bobcat 154
bobolink 314
Bombina
   orientalis 413
   variegata 412
Bombycilla garrulus 320
Bombyliidae family 539
Bonasa umbellus 277
bongo 194
bonobo 119
bontebok 195
bonytongues 454–5
Bos grunniens 198
Botaurus stellaris 254
Bothrops alternatus 394
Bothus lunatus 485
bovids 192–3
bowerbirds 334–5
bowfin 427, 448, 452
Brachiosaurus 47, 55
Brachycephalus ephippium 420
Brachygobius doriae 484
Brachylagus idahoensis 230
Brachylophus vitiensis 368
Brachyteles arachnoides 116
Bradypodion damaranum 367
Bradypus
   torquatus 92
   tridactylus 92
Branta
   leucopsis 260
   ruficollis 262
Brentidae family 532
bristlemouth 468
broadbill, green 316
bromeliads 16–17, 401
Bubalus quarlesi 198
Bubo virginianus 297
Bubulcus ibis 256
Bucephala clangula 263
budgerigar 291, 292
buffalo 198

Bufo
   pardalis 416
   viridis 417
bugs 490, 503, 518, 526–9
   jester 528
   peanut 528
bulbul
   red-vented 320
   white-cheeked 321
bullfinch, Eurasian 331
Bunolagus monticularis 230
bunting, painted 328
Buprestidae family 533
turbot, circumpolar 471
bustard
   black 279
   white-quilled 281
Buteo lagopus 268
Butorides
   striatus 256
   virescens 256
butterflies 491, 502, 518, 540–3
   bordered patch 543
   88 (eighty-eight) 543
   hackberry 542
   Malayan lacewing 543
   painted lady 542
   swallowtail 542
   white admiral 543
butterflyfish 427, 454
   Meyer's 481
buzzard, rough-legged 264

C

Cacajao calvus 117
Cacatua moluccensis 293
cactus, beaver-tail 21
caecilians 401, 408–9
   aquatic 409
   banna 401, 409, 583

   cayenne 409
   ringed 408
   Sao Tomé 409
   Southeast Asian 409
   terrestrial 401, 409
Caiman crocodilus 356
caimans 356–7
   black 357
   Cuvier's dwarf 357
   Nile 359
   Orinoco 359
   Schneider's dwarf 357
cale, rainbow 483
Callicebus personatus 117
Callipepla
   californica 276
   gambelii 277
Caloenas nicobarica 289
Caluromys derbianus 89
Calypte
   anna 303
   costae 303
Calyptomena viridis 316
Camarasaurus 54
camelids, New World 184–5
camels 182–3
   bactrian 183
Camelus
   bactrianus 183
   dromedarius 183
camouflage 442, 471
Camptosaurus 43
Canis
   aureus 128
   latrans 129
   lupus 127
   lupus dingo 127
   rufus 127
Caperea marginata 168
Capra aegagrus 199
Caprimulgus asiaticus 299
Caprolagus hispidus 231
capuchin, brown 116

capybara 216, 226
Carabidae family 533
caracal 155
caracara
　crested 264, 271
　striated 270
Caracara plancus 270
Carcharhinus melanopterus 438
Carcharias taurus 441
Carcharodon carcharias 441
Carcharodontosaurus 56
cardinals 329
Carduelis
　carduelis 331
　chloris 331
caribou 188, 190
Carnegiella strigata 463
carnivores 122–3, 142–3
cassowary 245
Castor
　canadensis 222
　fiber 223
Catagonus wagneri 181
caterpillars 541, 545
catfish 460–3
　electric 461
　striped sea 426, 462
cats 148–57
　African golden 157
　Andean mountain 156
　Asiatic golden 148, 157
　fishing 156
　jungle 156
　Pallas' 148
cattle 23, 178, 198–9
Cavia aperea 227
Cebus apella 116
centipedes 491, 503, 514–17
　giant desert 514
Centrophorus moluccensis 436
Centrosaurus 65
Centurio senex 102
century plant flowers 21

Cephalophus sp. 194
Cephalorhynchus
　commersonii 175
　hectori 174
Cerambycidae family 532
Ceratophrys ornata 415
ceratopians 64–5
Ceratosaurus 43
Ceratotherium simum 201
Cercopithecus diana 113
Cerocebus agilis 112
Cervus
　duvaucelii 190
　elaphus canadensis 191
Cestoidea class 498
Cetorhinus maximus 441
Chaetodipus formosus 225
Chaetodon meyeri 481
chain-fruit cholla 21
Chamaeleo
　jacksonii 365
　lateralis 367
chameleons 366
　Jackson's 362, 365
　Knysna dwarf 367
　Malagasy 367
characin, flying 427
Charadrius vociferus 284
Chasmosaurus 64
Chauliodus macouni 469
Chauna chavaria 262
cheetah 148, 155
Cheirolepsis 431
chimaeras 444–5
　blunt-nosed 446
　shortnose 445
chimpanzee 104, 119
chinchilla 227
chipmunk
　eastern 219
　least 217, 220
Chirocentrus dorab 459
Chironectes minimus 88

Chitala chitala 455
Chlamydosaurus knigii 367
Chlamydoselachus anguineus 443
Chloebia gouldiae 330
Chlorocebus aethiops 113
Chloropsis
　aurofrons 321
　hardwickii 321
Choerodon fasciatus 482
Chondrohierax urcinatus 267
chordates 492–3
chough 336
Chrysemys picta belli 352
Chrysococcyx cupreus 294
Chrysocyon brachyurus 127
Chrysomelidae family 532
cicada 502, 518, 526–7
cichlid, Lake Malawi zebra 483
Cicinnurus regius 334
Cidaris cidaris 500
Cinnyris regius 327
Cisco 464
Cissa chinensis 332
civet, African 122, 123, 142–3
Civettictis civetta 143
Cladoselache 431
classification 78–9, 236–7, 342–3,
　400–1, 490–1, 554–81
claws 85, 107, 148, 265
Cleidopus gloriamaris 477
Clethrionomys glareolus 225
Climatius 430
clingfish 474–5
clouded leopard 152
Clupeonella cultriventris 458
Cnemidophorus
　lemniscatus 377
　tesselatus 376
cnidarians 492–3
coati, South American 145
cobras 142, 390–1
　Egyptian 380, 392
　spitting 362, 390
Coccinellidae family 534

cock-of-the-rock, Guiana, 314, 317
cockatiel 292
cockatoos 290–3
　salmon-crested 293
　sulfur-crested 237
cockchafer 502
cockroaches 20, 43, 490, 518, 522–3
coconut 28
cod 448, 470
　Atlantic 426, 470, 582
coelacanth 451
Coelophysis 47
Coelurosaurus 346
Coelurus 43, 47
Coendou prehensilis 226
Colaptes auratus 310
Colius striatus 305
colobus 112–13
Colobus
　guereza 112
　satanas 113
Columba livia 289
Compsognathus 57
condor, Andean 264, 272
Condylura cristata 95
Conepatus
　chinga 139
　humboldtii 139
conifers 15, 24, 26
Connochaetus taurinus 194
Conraua goliath 419
coot, Eurasian 280
Copsychus malabaricus 322
Coracias caudata 309
Coracina lineata 320
coral reef ecosystem 492
coral 490, 493
Coregonus artedi 464
cormorants 252–3
　Brandt's 253
　double-crested 253

coruro 227
Corvus
　brachyrhynchos 336
　frugilegus 336
　monedula 337
Coryphaena hippurus 480
Corythaeola cristata 294
Corythosaurus 45, 60
coyote 129
crabs 30, 490, 515–17
　hermit 28, 503, 516
　purple shore 516
crake, red-necked 278
cranes 278–81
　black-crowned 237, 279
　red-crowned 280
　Siberian 278
crickets 524–5
　field 524
　great green bush 524
　Jerusalem 524
crocodiles 73, 358–9, 582–3
　African slender-snouted 358
　Cuban 358
　Siamese 358
crocodilians 354–9
Crocodylus
　cataphractus 358
　intermedius 359
　niloticus 359
　rhombifer 358
　siamensis 358
Crocuta crocuta 122, 123, 144
Crotalus polystictus 395
Crotaphytus collaris 369
crows 336–7
Crusafontia 45
crustaceans 514–17
Cryptobranchus alleganiensis 407
Cryptoprocta ferox 143
Ctenophorus isolepis gularis 367
cuatro ojos 475

cuckoos 23, 294–5
　African emerald 294
　channel-billed 295
　common 295
cuckooshrike, yellow-eyed 320
Cuculus canorus 295
Culex sp. 538
Culicidae family 538
curlew, Far Eastern 283, 285
cuscus, spotted 88
Cyanerpes cyaneus 328
Cyanistes caeruleus 326
cyclad 14
Cyclodomorphus gerrardii 377
Cyclopterus lumpus 479
Cygnus olor 263
Cylindrophus
　maculatus 383
　rufus 383
Cynocephalus
　variegatus 96
　volans 96
Cynognathus 82
Cyrtodactylus
　louisiadensis 372
　pulchellus 371

D

Dacelo sp. 306
　tyro 309
Dactylopsila trivirgata 84
Dallia pectoralis 467
Dama dama 191
Damasciscus pygargus 195
danio, zebra 427, 463
Danio rerio 463
Daption capense 250
Dasyprocta leporina 226
Dasyurus maculatus 89
Daubentonia madagascariensis 109

Decapoda order 516–17
deer 188–91
    fallow 79, 178, 191
Deinonychus 46, 346
Deinosuchus 67, 347
Delphinapterus leucas 173
Delphinus
    capensis 175
    delphis 174
Dendroaspis polyepsis 393
Dendrobates
    leucomelas 418
    pumilio 419
Dendroica pensylvanica 328
Dendrolagus goodfellowi 89
Dermogenys pusilla 474
Dermophis mexicanus 409
Deroptyus acciptrinus 292
deserts 20–1
desman, Pyrenean 94
Desmodus rotundus 102
Diademichthys lineatus 474
Diaethria sp. 543
Dicamptodon ensatus 404
Dicerorhinus sumatrensis 201
Diceros bicornis 201
Dicrurus paradiseus 332
Didemnum molle 493
Dimetrodon 38, 66, 82, 346
Dimorphodon 66
dingo 127
Dinopium benghalense 313
Dinornis maximus 241
dinosaurs 37–9, 46–7, 72–3
    anatomy 48–51
    armored 62–3
    contemporaries 66–7
    Cretaceous 44–5
    eggs 52–3
    extinction 68–9
    fossils 70–1
    Jurassic 42–3
    raising young 52–3

    Triassic 40–1
Diogenidae 516
Diplodactylus ciliaris 370
Diplodocus 43, 54
Diplopoda class 516
Dipodomys deserti 225
Diporiphora superba 367
dogfish
    pike 434
    prickly 437
    spiny 436
dogs
    bush 127
    domestic 125–6
    raccoon 126
    wild 124–7
Dolichotis patagonium 227
dollarbird 309
dolphinfish, common 480
dolphins 162–3, 170, 174–7
    Amazon River 162, 176
    Atlantic white-sided 164
    bottlenose 164, 175
    Commerson's 175
    Ganges River 162, 176
    Hector's 174
    Indus River 176
    long-beaked 175
    long-snouted spinner 175
    vaquita 177
    white-beaked 174
    Yangtze River 162, 176
dormouse
    desert 224
    masked mouse-tailed 224
    woodland 224
Dorosoma cepedianum 458
dorsal-band whale 454
Dorudon 67
Doryrhamphus dactyliophorus 478
douc langur 114
doves 288–9

dragonfish 426, 448, 468–9
dragonfly 38, 41, 490, 503, 518, 520–1
Drepanaspis 39
drill 113
dromedary 178, 179, 183
drongo, greater racket-tailed 332
Dryocopus pileatus 310
ducks 260–3
    American black 263
    wood 262
dugong 30, 212
Dugong dugon 212
duiker 194
Dunkleosteus 38, 430
dunnock 323
Dytiscidae family 533

E

eagles 264–7
    Bonellis 236
    crested serpent-eagle 267
    great Philippine 267
    harpy 265
    ornate hawk 269
    white-bellied sea 264, 266
ears 81, 99
earthworms 491, 498
Echeneis naucrates 480
echidna 81
    long-beaked 84
    short-beaked
echinoderms 500–1
Echinorhinus brucus 437
Echinosorex gymnura 94
echolocation 99, 170
ecosystems 12–13, 492
Ectophylla alba 103
edelweiss 26
Edmontonia 63, 343

Edmontosaurus 49, 50
eels 456–7, 582
    American 456
    Congo 401, 402, 406
    ribbon 427, 448, 456
    tessellated moray 456
    umbrella mouth gulper 457
    white spotted spiny 472, 478
egret
    cattle 257
    great white 254
    little 254
Egretta
    alba 254
    caerulea 256
    garzetta 254
eider 263
    king 262
Elanoides forficatus 266
Elanus caeruleus 267
Elaphe mandarina 388
Elasmosaurus 347
electrosense 434
elephantnose 455
elephants 208–11
    African 79, 208–11
    Asian 208, 210–11
Elephantulus
    myurus 214
    rozeti 215
    rufescens 215
Elephas
    maximus indicus 211
    maximus maximus 211
    maximus sumatranus 210
Eleutherodactylus augusti 414
Empidae family 539
emu 73, 244–5
Emydocephalus annulatus 393
Engraulis ringens 458
Ephippiorynchus senegalensis 254
Epicrates cenchria cenchria 385
Epinephelus lanceolatus 481

epiphytic plant 16
Epomorphorus gambianus 100
Eptatretus stouti 432
Equus sp. 205
    asinus 205
    caballus 205
    caballus gmelini 205
    kiang 205
    onager 205
Eretmochelys imbricata bissa 353
Erinaceus europaeus 94
ermine 141
Eschrictius robustus 168
Esox lucius 466
Etmopterus lucifer 437
Eubalaena australis 169
Eublepharus macularius 371
eucalyptus flower 18
Euderma maculatum 102
Eudimorphodon 41
Eumeces fasciatus 379
Euoplocephalus 45, 47, 63
Euphausiidae family 516
Eupholus bennetti 532
Euplectes orix 330
Eupodotis
    afra 279
    afranaoides 281
Eurostopodus argus 298
Eurpyga helias 279
Eurycea bislineata 406
Eurypharynx pelecanoides 457
Eurystomus orientalis 308
Eusthenopterion 431
evolution 14–15, 82–3, 104,
    122, 130, 148, 162, 178,
    216, 240–1, 264, 314,
    346–7, 362, 380, 402,
    410, 430–1, 434, 448,
    472, 504, 518
    convergent 81
Exocoetidae family 474
eyes 85, 182

F

Facipennis canadensis 277
fairywren, splendid 314, 325
Falco
    peregrinus 271
    rupicoloides 270
    sparverius 270
    subbuteo 271
falconet, spot-winged 270
falcons 270–1
    peregrine 264, 271
    pygmy 271
Falcunculus frontatus 324
fangs *see* teeth
featherback, Asiatic 455
feet and hands 54, 58, 104, 130,
    208, 245, 362
Felis
    chaus 156
    jacobita 156
    wiedii 157
figbird, green 332
finches 330–1
    Galapagos 315
    golden 330
    Gouldian 330, 580
firecrest 324
fireflies 533
fish 73, 582
    bony 428, 448–9
    characteristics 428–9
    classifying 426–7, 564–72
    evolution 430–1
    jawless 432–3
    spiny-rayed 472–3
flameback, black-rumped 313
flamingos 258–9, 580
flashlight fish 476
flatfish 484–5
fleas 518, 536–9
flies 502, 518, 536–9
    caddis 536

deer 536
fruit 539
hairy bee 539
house 538–9
tsetse 539
flight 242–3
flounder 472, 485
flying fish 474
flying fox
   gray-headed 100
   Indian 101
Fordonia leucobalia 386
Formicidae family 546, 549
fossa 143
fossils 70–1
fox
   Arctic 128
   bat-eared 128
   gray 129
   kit 129
   red 124
Fratercula arctica 287
frogmouths 298–9
frogs 400, 410–21, 583
   Asian painted 420
   brown New Zealand 400, 410, 413
   bull 414, 415
   cape clawed 413
   cape ghost 420
   common parsley 400, 410, 413
   corroboree 415
   Darwin's 421
   European painted 400, 410, 413
   funereal poison 418
   glass 417
   gold 420
   goliath 419
   green and golden bell 416
   leopard 400
   Mexican burrowing 400, 410, 412
   New Zealand 413
   Northern leopard 418
   orange and black poison 418
   ornate burrowing 418
   painted reed 421
   paradox 421
   pickeral 419
   red-banded crevice creeper 421
   Schmidt's forest 415
   Senegal running 420
   Seychelles 420
   Southern platypus 414
   spotted grass 415
   strawberry poison 400, 410, 419
   tailed 400, 410, 412
   tree 410, 417, 418
   variable harlequin 416
   Wallace's flying 400, 421
   Western barking 414
   Western marsh 414
Fulgora sp. 528
Fulica atra 280

G

Gadus morhua 470
Galago senegalensis 106
Galbula ruficauda 313
Galemys pyrenaicus 94
Gallinago hardwickii 284
Gallinula chloropus 260, 281
gallinule, purple 279
Gallirallus
   australis 280
   philippensis 280
Gallotia Atlantica 378
Gallimimus 57
gamebirds 274–7
Gampsonyx swainsonii 268
gannet 252

gar, spotted 427, 448, 452
Garrulus glandarius 337
Gasteracantha sp. 511
Gastrophryne carolinensis 421
gaur 193
Gavia sp.
   immer 248
   stellata 249
Gazella thomsonii 195
gazelle, Thomson's 195
Gecko gecko 370
geckos 20, 342, 370–3
   blue-tailed day 372
   Cogger's velvet 370
   common wonder 371
   flap-footed 370
   flying 362, 372
   Gray's bow-fingered 371
   Israeli fan-fingered 373
   leopard 342, 371
   mourning 373
   Northern spiny-tailed 370
   ring-tailed 372
   Southern spotted velvet 373
   Tokay 370
   yellow-headed 371
geese 260–1
   barnacle 260
   graylag 262
   perigord 236
   red-breasted 262
   snow 241, 263
gemsbok 78, 195
Geochelone sp. 351
Geococcyx californianus 295
gerbil 225
Gerbillus sp. 225
gerenuk 192, 195
Geronticus eremita 256
gharials 358–9
gibbon, lar 104, 105, 111
Gigantactis paxtoni 471
gila monster 375

ginger, flowering 16
ginko 14
giraffes 79, 178, 186–7
Glaucomys volans 221
gliding 96
Glossina sp. 538
Gnathonemus petersii 455
goanna 374
goats 193, 198–9
   mountain 199
   wild 199
gobies 484–5
   bluebanded 484
   decorated fire 484
   Doria's bumblebee 484
godwit, Hudsonian 285
goldcrest 324
goldeneyes, common 263
goldfinch 314, 331
Gonatodes albogularis fuscus 371
Gonostomatidae family 468
gorillas 78, 104, 118, 120–1, 581
goshawk, eastern chanting 269
gourami, splendid licorice 480
Grammistes sexlineatus 480
Graphiurus murinus 224
grasshoppers 490, 502, 518, 524–5
grayling 466
greater roadrunner 237, 295
grebes 248–9
   eared 248
   great crested 248
   horned 237, 248
   little 249
   pied-billed 248
   red-necked 248
green-billed malkoha 294
greenfinch, European 331
ground-roller 307
groupers 480–1
   giant 481

grouse 24, 275
   ruffed 277
   spruce 277
Grus japonensis 280
Gryllacrididae family 524
Gryllidae family 524
guan, white-crested 276
guanaco 185
guinea pig 227
guineafowl, vulturine 277
gulls 237, 286–7
Gymnothorax favagineus 456
Gypaetus barbatus 273
Gypohierax anglolensis 272

H

habitats 14–15
hagfish 432–3
   Atlantic 432
   Pacific 433
hair, mammals 81
halfbeak, Malayan 474
Haliaeetus leucogaster 266
Haliastur indus 266
Haramiya 41
hares 228–31
   Arctic 229
   Central African 231
   hispid 231
   snowshoe 229
Harpactes erythrocephalus 305
hatchetfish 469
   marbled 463
hawks 268–9
   African harrier 237, 269
   rough-legged 268
   slate-colored 268
headstander, striped 463
hedgehog, Western European 94
Helarctos malayanus 130, 137

Heleiporus barycragus 414
Heleophryne purcelli 420
heliobatus ray 37
hellbender 401, 402, 407
Heloderma suspectum 375
Hemicentetes semispinosus 95
Hemicyclapsis 430
Hemigrapsus nudus 516
Hemiprocne longipennis 302
Hemiscyllium ocellatum 443
Hemitragus jemlahicus 199
Heosemys spinosa 352
hermit crab 503, 516
herons 254–7
   black-crowned 254
   goliath 257
   gray 256
   green 256
   little blue 256
   purple 237, 257
Herpailurus yagouaroundi 157
Herrerasaurus 41
herring
   Atlantic thread 458
   blackfin wolf 459
   Pacific 427
Hesperornis 240
Heterodontosaurus 43
Heterodontus portusjacksoni 443
Heterohyrax brucei 206
Heteroptera family 528
Hexanchus griseus 443
Hexaprotodon liberiensis 180
Hexatrygon bickelli 446
Hildebrandtia ornata 418
Hiodon tergisus 455
Hippocampus abdominalis 478
hippopotamus 162, 178, 180, 581
   pygmy 180
Hippopotamus amphibius 180
Hippotragus niger 194
hoatzin 276
hobby, Eurasian 271

hogs 181
Homaloptera orthogoniata 461
Homoroselaps lacteus 388
honeycreeper, red-legged 328
honeyeater
 banded 314
 regent 326
hoopoe 307, 580
hornbill, yellow-billed 309
horns 178
horses 79, 178, 202–5
horsetail fern 14
hound, pharoah 123
hummingbirds 236, 300–3
 Anna's 301, 303
 black-chinned 301, 302
 blue-throated 303
 broad-tailed 303
 Calliope 303
 Costa's 301, 303
 ruby-throated 301, 302, 580
 rufous 301, 302
hunting 128, 166, 253
Huso huso 453
Hydrochaeris hydrochaeris 226
Hydrolaetare schmidti 415
Hydrolagus colliei 446
Hydrosaurus ambionensis 366
Hyemoschus aquaticus 190
hyena 146–7
 spotted 122, 123, 144, 148
 striped 144
Hylinobatrachium fleischmanni 417
Hylonomus 38, 346
Hymenopodidae family 521
Hyperolius marmoratus 421
Hyperoodon ampullatus 173
Hypsignathus monstrosus 100
Hypsilophodon 46
Hypsipyrmnodon moschatus 90
hyraxes 27, 79, 206
 rock 27, 206
 yellow-spotted 206

Hystix
 brachyura 226
 cristata 226

I

Ichneumonidae family 548
Ichthyophis
 bannanicus 409
 kohtaoensis 409
Ichthyornis 240
Ichthyosaurus 67, 346–7
Ichthyostega 38
Icterus
 auratus 333
 bullockii 332
 nigrogularis 333
Idiacanthus fasciola 469
iguana 368–9
 Fijian crested 368
 helmeted 364
 marine 342, 362, 365, 369
iguanids 364–5
Iguanodon 61
Imantodes cenchoa 388
indri 106
Indricotherium 83
Inia geoffrensis 176
invertebrates
 classifying 490–1, 573–81
Isistius brasiliensis 436
Isurus oxyrinchus 441

J

jacamar, rufous-tailed 313
jackal, golden 128
jackdaw, Eurasian 337
Jaculus sp. 225

jacuna 283
jaguar 148, 153
jaguarundi 157
jays 336–7
jellyfish 38, 491, 493
jerboa 216, 225
julie, striped
jungle runner 379, 580

K

kagu 281
kakapo 291
Kaloula pulchra 420
kangaroos 87, 90–1
 goodfellow's tree 90
 musky rat 90
 red 87, 581
 western grey 90
Kannemeyeria 41
Kassina senegalensis 420
kea 291
Kentropix calcarata 377
kestrel
 American 270
 greater 270
 lesser 264
kiang 205
killdeer 284
killifish
 clown 472, 475
 steel-blue 475
kingbirds 316–17
kingfishers 306–9
 belted 308
 malachite 306
 white-throated 236, 306
kinglet, golden-crowned 325
kinkajou 144
kiskadee, great 236, 314, 316

kites 264, 266–7
    black-shouldered 267
    brahminy 266
    hook-billed 267
    pearl 268
    slate-colored 267
    snail 267
    swallow-tailed 266
kiwi 236, 244–5
klipspringer 194
knifefish, South American 426
koala 81, 89
kodkod 156
Kogia breviceps 173
komodo dragon 374
kookaburra 306–7, 309
krill 516
Kronosaurus 45
kudu, greater 193

L

Lacerta vivipara 376
lacewing 503, 527
ladybugs 502, 534–5
Lagenorhynchus albirostris 174
Lagostomus maximus 227
Lama
    glama 185
    guanicoe 185
    pacos 184
Lambeosaurus 60
lammergeier 273
Lamna
    ditropis 440
    nasus 441
Lampetra
    fluvistilis
    planeri 432
Lampornis clemenciae 303
lamprey
    European brook 426, 432

European river 432
    sea 432
Lampris guttatus 477
Lampropeltis getulus 387
Lamprophis guttatus 387
Lamprotornis superbus 333
Lampyridae family 533
Lanius schach 321
lanternfish 448, 468–9
lapwing, banded 284
Larus
    argentatus 287
    cachinnans 287
    canus 287
    delawarensis 287
Laticauda colubrina 392
Latimeria chalumnae 451
Lavia frons 102
leafbird
    golden-fronted 321
    orange-bellied 314, 321
leafhopper 526–7
leeches 498
Leiarius sp. 462
Leiopelma hamiltoni 413
Leiopelmatidae family 413
leiothrix, red-billed 322
Leiothrix lutea 322
lemming 224
Lemmus sp. 224
lemurs 79
    brown mouse 109
    Coquerel's mouse 109
    flying 96
    fork-marked 108
    gray-mouse 106
    red ruffed 108
    ring-tailed 104, 105, 107
leopard 148, 152–3 *see also*
        clouded leopard, snow
        leopard
    cat 156
Lepidodactylus lugubris 373
Lepidostren paradoxa 450

Lepisosteus oculatus 452
Leptailurus serval 157
Leptodactylus pentadactylus
    415
Leptotyphlops humilis 382
Leucopternis schistacea 268
lice 522–3
limbs 349 *see also* feet and
    hands
Limnodynastes
    interioris 414
    tasmaniensis 415
Limosa haemastica 285
limpkin 278
linsang, banded 143
lion 122, 148, 150–1
Liopleurodon 67
Lipophrys nigriceps 482
Lipotes vexillifer 176
Litocranius walleri 195
Litoria
    aurea 416
    gracilenta 417
lizardfish, variegated 426, 448,
    468
lizards 73, 342–3, 362–79
    basilisk 369
    bearded 374–5
    checkered whiptail 376
    collared 364, 369
    desert spiny 368
    dwarf flat 378
    frill-necked 342, 362, 367
    granite night 377
    haria 378
    Italian wall 343, 379
    legless 342, 362
    rainbow 377
    salt-tailed water 366
    skinklike 376–9
    South American skinklike 377
    viviparous 376
llama 184–5

loach
    saddled hellstream 461
    slimy 460
lobster 490, 515
    spiny 514–15
locust 524
logrunner 324
Lonchorhina sp. 103
loons 248–9
    red-throated 237, 249
Lophelia pertusa 493
Lophiomys imhausi 224
loris, slow 109
Lorius lory 293
lory, black-capped 291, 293
Lota lota 471
louse 522
lovebird 291
Loxocemus bicolor 383
Loxodonta
    africana 210
    cyclotis 211
Luciocephalus pulcher 481
lumpsucker 479
lungfish
    African 451
    Australian 450
    South American 427, 450
Lutra lutra 140
Lybius bidentatus 313
Lycaon pictus 127
Lymantridae family 545
lynx 148, 154–5
Lynx
    canadensis 154
    lynx 154
    pardinus 155
    rufous 154
lyrebird, superb 314
Lythrypnus dalli 484

M

Macaca
    arctoides 115
    nemestrina 115
    nigra 114
macaque
    Celebes crested 114
    pigtail 115
    stump-tailed 115
macaw, scarlet 291
Macrocephalon maleo 274
Macronectes giganteus 250
Macropus sp. 90
    fuliginosus 90
    rufus 87
Macroscelides proboscideus 215
Macrotis lagotis 89
magnolia 15, 45
magpie 236, 337
magpie, green 332
Maiasaura 47
Malapterurus electricans 461
Malayemys subtrijuga 352
maleo 274
malleefowl 274
Malurus splendens 325
mamba, black 393
mammals 37, 69
    characteristics 80–1
    classifying 78–9, 555–7
    evolution 82–3
    hoofed 178–205
    insect-eating 94–5
mammoth, woolly 83
manakin, wire-tailed 317
manatees 79, 212–13
mandarin fish 474
mandrill 105
Mandrillus leucophaeus 113
mangabey, agile 112
Mantidae family 521

mantids 502, 518, 520
    orchid 521
    praying 521
maple seeds 24
mara, Patagonian 227
marbled cat 152
margay 148, 157
marmot 219, 220
Marmota sp. 219
    monax 219
marsupials 37, 86–9
marten 140
    European pine 141
    yellow-throated 141
Martes
    flavigula 141
    martes 141
martins 319
Mastacembelus armatus 478
Maticora bivirgata 392
mayfly 490, 503, 518, 520–1
meerkat 143
Megalosaurus 58
Megaptera novaeangliae 169
Megazostrodon 82
Megophrys nasuta 412
Melamosuchus niger 357
Melanorosaurus 41
Meles meles 139
Melierax poliopterus 269
Melopsittacus undulatus 292
Melursus ursinus 130, 136
Merops
    nubicus 308
    ornatus 308
Mesonyx 66
Microcebus
    coquereli 109
    murinus 106
    rufus 109
Micruroides euryxanthus 392
military dragon 367
milkfish 427, 460

millipede 490, 516
   flattened 514
minivet, scarlet 320–1
mites 504, 512–13
Mitsukurina owstoni 441
mockingbird 323
Mola mola 484
moles 94–5
   European 94, 95
   star-nosed 95
mollusks 490, 494–5
Moloch horridus 366
Momotus momota 306
monarch, spectacled 325
Monarcha trivigatus 325
mongoose 142
   dwarf 122
monitor lizard 343, 362, 583
   crocodile 375
   emerald tree 374
monkeys *see also* apes, primates
   African Old World 112–13
   Asian Old World 114–15
   Allen's swamp 112
   Diana 113
   dusky leaf 115
   golden snub-nosed 104, 115
   New World 116–17
   northern night 116
   proboscis 105, 114
   red howler 116
   spider 78, 104, 111
   vervet 110, 113
   woolly 110
Monodon monoceros 173
monotremes 84–5
mooneye 455
moonrat 79, 94
moorhens 260, 281
Moorish idol 481
moose 189, 191
Morelia spilota 385
Morganucodontid 82

mosquitos 502, 538–9
   anopheles 538
   culex 538
Motacilla sp. 318
   cinerea 318
moths 540–1, 544–5
   Cerisy's sphinx 545
   five-spotted burnet 545
   Hercules 541
   hummingbird 544
   Indian moon 541, 544
   Madagascan sunset 545
   white-lined sphinx 544
   yellow emperor 545
motmots 306–7
mouflon 199
mouse
   grasshopper 225
   long-tailed pocket 225
   Mexican spiny pocket 217
mousebird, speckled 305
mountain anoa 198
mud whelk 495
mudpuppy 401, 402, 405
mudskippers 482
mullet 472
Muntiacus muntjak 190
muntjac, Indian 178, 190
muriqui 116
Muscidae family 538–9
muskox 199
muskrat 224
mussels 28, 495
Mustela
   erminea 141
   lutreola 141
   nivalis 141
   putorius 138
mustelids 138–41
Mycteria leucocephala 256
Myctophidae family 468
Myomimus personatus 224
Myrmecobius fasciatus 86

Myrmecophaga tridactyla 93
Myxine glutinosa 432

N

Naja haje 392
narwhal 162, 171, 173
Nasalis larvatus 114
Natrix natrix 387
Nausa nausa 145
nautilis 494
Necturus masculosus 405
Nematoda phylum 498
Neoceratodus fosteri 450
Neofelis nebulosa 152
Neophocaena phocaenoides 177
Nephila sp. 511
Nesoclopeus woodfordi 278
Nesolagus netscheri 230
Netmateleotris decora 484
newts 402–7
   alpine 407
   marbled 406
nightjars 236, 298–9
niltava, rufous-bellied 325
Niltava sundara 325
Noctilio leporinus 103
northern flickers 312
northern screamer 262
noses 94, 99, 355
Notaden bennettii 414
Nothosaurus 41, 67
Notodontidae family 545
Notoraja sp. 446
Nudibranchia order 494
numbat 86
Numenius madagascariensis 285
nuthatch
   Eurasian 327
   white-breasted 327

Nyctereutes procyonoides 126
Nyctibius griseus 298
Nycticebus coucang 109
Nycticorax nycticorax 254
Nyctidromus albicollis 299
Nymphalidae family 542–3
Nymphicus hollandicus 292

O

oarfish 476
oceans 30–1
ocelot 123, 148, 150
Ochotona
   alpina 231
   curzoniae 228
   princeps 231
   roylei 231
octopus 496
Oedura
   coggeri 370
   tryoni 373
oilbird 237, 299
okapi 187
Okapia johnstoni 187
Oligochaeta class 498
olm 405
Ommastrephes sagittatus 496
onager 205
Oncifelis guigna 156
Oncorhynchus
   aguabonita 466
   mykiss 466
   nerka 467
Ondrata zibethicus 224
Onychodactylus fischeri 404
Onychomys sp. 225
opah 426, 448, 477
Ophisaurus apodus 376
Opisthocomus hoazin 276
Opisthonema oglinum 458

opossum
   water 88
   woolly 89
orangutan 78, 104, 119
Orcinus orca 174
Oreamnos americanus 199
Orectolobus ornatus 443
oreo, spotted, 472, 477
Oreotragus oreotragus 194
oriole
   bullock's 332
   Eurasian golden 314, 332
   orange 333
   yellow 333
Oriolus oriolus 332
Ornithomimus 45
ornithopods 60–1
Ornithorhynchus anatinus 85
Ornithosuchus 38
Orthonyx temminckii 324
Orycteropus afer 206
Oryx gazella 195
Oryzias javaniaus 475
oscar 483
osprey 264, 265, 268
Osteoglossum bicirrhosum 454
ostrich 245, 580
Otocyon megalotis 128
otter, 122
   European 140
   giant 140
   sea 140
Ouranosaurus 46, 60
Ovibos moschatus 199
Oviraptor 52–3, 57, 59
Ovis
   canadensis 199
   orientalis musimon 199
owls 24, 237, 264, 296–7
Oxybelis sp. 389
Oxynotus bruniensis 437
oystercatcher 282
oysters 495

P

Pachycephalosaurus 45, 47
Pachyrachis 347
Padda orzivora 330
paddlefish
   American 453
   Chinese 453
Paleosuchus
   palpebrosus, 357
   trigonatus 357
Palinurus vulgaris 514
Pan
   paniscus 119
   troglodytes 119
panda
   giant 130, 137
   red 130, 145
Pandion haliaetus 268
pangolins 78, 81, 92–3
panther 153
Panthera
   leo 151
   onca 153
   pardus 153
   tigris 151
Pantodon buchholzi 454
Papilionidae family 542
Papio hamadryas 113
Paradisaea apoda 334
Paradoxurus hermaphroditus 142
Paragalaxias mesotes 464
Parasaurolophus 46, 61
Pardofelis marmorata 152
Pardolatus puncatus 327
Parosphromenus dreissneri 480
parotia, Western 334
Parotia sefilata 334
parrots 290–3
   Pesquet's 293
   red-fan 292
partridge 275

Parus
    inomatus 327
    major 326
Passerina ciris 328
passerines 314–15
pauraque 299
Pavo cristatus 276
peafowl, Indian 276
peccary, Chacoan 178, 181
Pedionomus torquatus 281
Pedostibes
    everetti 416
    hosii 417
Pelamis platurus 393
Pelecanus
    erythorynchos 253
    occidentalis 252
pelicans 237, 252–3
Pelodytes punctatus 413
Peloneustes 67
Penelope pileata 276
penguins 246–7, 580
    adelie 237
    chinstrap 247
    emperor 246–7
    Fjordland 247
    magellanic 247
    rockhopper 247
    yellow-eyed 247
Pentalagus furnessi 230
perch 480–1
    cross-toothed 468
    pirate 470, 472
    yellow 472
perentie 375
Pericrocotus flammeus 320
Periparus ater 326
Petaurus norfolcensis 88
petrel
    Antarctic giant 250
    cape 250
Petrocephalus simus 454
Petrodromus tetradactylus 215

Petrogale xanthopus 91
Petroica goodenovii 324
Petromyzon marinus 432
Phacoboenus australis 270
Phaenicophaeus tristis 294
Phalacrocorax
    auritus 253
    penicillatus 253
Phaner furcifer 108
Phaocchoerus africanus 181
Pharomachrus mocinno 304
Pharscolarctos cinereus 89
Phataginus sp. 92
pheasant 275
    Reeve's 277
Phelsuma cepediana 372
Phocoena
    phocoena 177
    sinus 177
Phocoenoides dalli 177
phoebe, black 317
Phoenicopterus
    jamesi 258
    minor 258
Photoblepharon palpebratus 476
Phrynocephalus persicus 366
Phrynomerus bifasciatus 421
Phrynosoma douglassii 368
Physeter catodon 173
piculet, rufous 311
pigeons 237, 288–9
pigs 180–1
pikas 23, 228–31
    American 213
    black-lipped 228
    northern 231
    Royle's 213
pike, Northern 426, 448
pikehead, South-East Asia 481
pilchard, European 458
pimelodid, sailfin 462
pine 24, 28
pineapplefish, Australian 477

Pipa pipa 412
pipefish 478–9
pipit, water 318
Pipra filicauda 317
Piranga olivacea 328
piranha, red-bellied 448, 462
Piscicolidae family 498
Pitangus sulphuratus 316
Pithecophaga jefferyi 267
pitta, banded 316
Pitta guajana 316
plains viscacha 227
plains-wanderer 281
Planocephalosaurus 347
Platalea leucorodia 254
Platanista
    gangetica 176
    minor 176
Platecarpus 67
Platemys platycephala
    platycephala 350
Plateosaurus 47, 55
Platycercus eximus 293
Platymantis guppyi 418
platypus 78, 85
Platysaurus guttatus 378
Platysternon megacephalus 350
Pleisochelys 43
Plotosus lineatus 462
plover 283
    American golden 285
    black-bellied 284
Pluvialis
    dominica 285
    squatarola 284
Podarcis
    perspicillata 378
    sicula 379
Podargus strigoides 298
Podiceps
    auritus 248
    grisegena 248
    nigricollis 248

Podilymbus podiceps 248
Podocnemis unifilis 351
Poecilogale albinucha 141
Poelagus majorita 231
Pogonophora class 499
polecat
　European 138
　marbled 138
Polihierax semitorquatus 270
Polyboroides typus 269
Polydesmus complanatus 514
Polyglyphanodon 45
Polyodon spathula 453
Polypterus weeksi 452
Pongo pygmaeus 119
porcupine
　Brazilian 226
　crested 216, 226
　Malayan 226
　prehensile-tailed 217
Porphyrio mantelli 280
Porphyrula martinica 279
porpoises 162–3, 177
　Dall's 162, 177
　finless 177
　harbor 177
　spectacled 177
Portuguese man o'war 493
possum 81, 87
Potamochoerus porcus 181
Potamotrygon motoro 446
potoo, common 298
Potos flavus 144
potto, golden 108
prairie dogs 218
prairie-chicken, greater 276
primates 37, 104–5 *see also* apes, monkeys
　faces 111
　higher 110–11
　lower 106–9
Prionailurus
　bengalensis 156
　viverrinus 156

Prioncdon linsang 143
Prionodura newtoniana 335
Pristophorus cirratus 437
Procavia capensis 206
Procompsognathus 41
Procyon lotor 145
Profelis
　aurata 157
　temmincki 157
Proganochelys 346
pronghorn 178, 187
Propithecus
　diadema 108
　verreauxi 106
Proscyllium habereri 438
Proteles cristatus 145
Proteus anguinus 405
Protopterus sp. 451
Protoreaster linckii 500
Protriceratops 65
Psephurus gladius 453
Pseudanthis tuka 480
Pseudechis porphyriacus 393
Pseudis paradoxia 421
Pseudobranchus striatus 405
Pseudocyttus maculatus 477
Pseudophryne corroboree 415
Pseudorca crassidens 173
Pseudotriton ruber 407
Pseudotropheus zebra 483
Psittrichas fulgidus 293
ptarmigan 275
Pteranodon 346–7
Pterapsis 39
Pternohyla fodiens 417
Pterocles alchata 288
Pterodaustro 66
Pteroglossus beauharnaesii 313
Pteronura brasiliensis 140
Pteropus
　giganteus 101
　poliocephalus 100
Ptilocercus lowii 97

Ptilonorhynchus violaceus 335
Ptychozoon sp. 372
Ptyodactylus puiseuxi 373
pudu 190
puffbird, chestnut-capped 311
puffins 287
　horned 282
Puffinus sp. 250
Pulicidae family 538
puma 148, 151
Puma concolor 151
purplequeen 480
puya 26
Pycnonotus
　cafer 320
　leucogenys 321
Pygathrix
　nemaeus 114
　roxellana 115
Pygoncentrus nattereri 462
Pygopodidae, family 370
Pyrrhocorax
　graculus 336
　pyrrhocorax 336
Python curtus 385
pythons 384–5
　blood 385
　burrowing 380, 383
　carpet 380, 385
Pytilia melba 330

Q

quail
　California 276
　Gambel's 236, 277
quiver tree 21
quoll, spotted-tail 89

R

rabbits 78, 228–31
   antelope jackrabbit 229
   black-tailed jackrabbit 229
   brush 230
   pygmy 230
   rivenne 230
   ryuku 230
   volcano 230
raccoons 122, 123, 130, 144–5
rail
   buff-banded 280
   Woodford's 278
rainbow lorikeet 291, 292
rainbowfish
   Australian 473
   Madagascar 472, 475
rainforests, tropical 16–17
Raja sp. 446
Rallina tricolor 278
Ramphastos
   sulfuratos 312
   toco 310
Ramphotyphlops braminus 382
Rana sp. 418
   palustris 419
Rangifer tarandus 190
raptors 264–73
rats 216, 581
   black 225
   crested 224
   desert kangaroo 216, 225
   house 225
   spiny 227
rattlesnakes 390–1
   lance-headed 395
Rattus
   mindorensis 225
   rattus 225
Raturfa bicolor 220
raven 337

rays 444–7
   Atlantic devil 445
   blind electric 446
   electric 434, 445
   manta 446
   sting *see* stingrays
   torpedo 445
Regalecus glesne 476
Regulus
   ignicapillus 324
   regulus 324
   satrapa 325
reindeer 179
remora 480
reproduction 36–7, 80, 86, 132, 348, 435, 448, 479, 506
reptiles
   characteristics 344–5
   classifying 342–3, 561–2
   evolution 346–7
resplendent quetzals 236, 304, 305
Rhacophorus nigropalmatus 421
Rhamphorynchus 43
rhea 245
Rheobatrachus silus 414
Rhincodon typus 443
Rhinoceros
   sondaicus 201
   unicornis 201
Rhinoderma darwinii 421
Rhinolophus sp. 102
Rhinomuraena quaesita 456
Rhinophis drummondhayi 383
Rhinophrynus dorsalis 412
rhinoceros 79, 200–1
   white 178–9
Rhodeus sericeus 460
Rhyacotriton olympicus 406
Rhynchocyon chrysopygus 215
Rhynchops niger 285
Rhynochetos jubatus 281
ribbonfish 477

ricefish, Javanese 475
ringtail 145
robin 315, 322–4
rockfish, tiger 472, 479
rodents 216–27
roller, lilac-breasted 309
Romerolagus diazi 230
rosella, Eastern 291, 293
Rostrhamus sociabilis 267
roughy, Southern 472, 477
Rousettus egyptiacus 101
Rupicola rupicola 317

S

Sagittarius serpentarius 272
Saguinus
   midas 116
   oedipus 117
Saiga tatarica 195
salamanders 402–7
   blue-spotted 404
   fire 402–4
   Fischer's clawed 401, 402, 404
   Japanese giant 407, 583
   Olympic torrent 401, 402, 406
   Pacific giant 401, 402
   red 401, 402, 407
   tiger 401, 402, 405
   two-lined 406
Salamandra salamandra 404
Salmo salar 467
salmon 464–7
   Atlantic 426, 467
   sockeye 464–5, 467
Saltasaurus 45, 47
Saltopus 41
Salvelinus namaycush 466
sandgrouse 288–9
sandpiper, spotted 283, 284
Sanopus splendidus 470

Sarcophilus laniarius 84
Sardina pilchardus 458
Sardinella aurita 459
sardines 448, 458–9
   Spanish 459
Saturniidae family 541, 545
sauropods 54–5
savannas 18–19
Sayornis nigricans 317
scales 344, 429
Scaphiopus couchii 413
Sceloporus magister 368
scheltopusik 343, 376
Schistometopum thomense 409
Sciurus sp. 219
   aberti 219
   niger 219
   niger cenerus 220
   vulgaris 221
Scolopendra heros 514
scorpion 39, 491, 504, 512–13, 518
Scythrops novaehollandiae 295
sea anemone 28, 490, 493
sea cucumber 501
sea krait, yellow-lipped 392
sea slug 494
sea squirt 493
   colonial 493
sea urchin 500
seabass 480–1
seahorses 426, 448, 472, 478–9
sealion 78, 122, 158–61
   steller 160
seals 158–61
   baikal 161
   bearded 160
   bull elephant 161
   crabeater 161
   harbor 161
   harp 122, 160, 161
   hooded 161
   leopard 161
   New Zealand fur 160, 161
   ribbon 161
   weddell 161
seashores 28–9
Sebastes nigrocinctus 479
secretary bird 264, 272
Selasphorus
   platycercus 303
   rufus 302
Selevinia betpakdalaensis 224
Sellosaurus 41
serval 157
shad, American 458–9
shag 252
shama, white-rumped 322
sharkminnow, tricolor 461
sharks 428–9, 434–43
   basking 441
   blackbelly lantern 437
   blacktip reef 438
   blue 444
   bluntnose sixgill 443
   bramble 437
   broadnosed 7-gill 434
   bulltip 427
   Californian horn 434
   common saw 434
   cookie-cutter 436
   crocodile 434
   epaulette 443
   frilled 443
   goblin 441
   graceful cat 438
   great white 427, 441
   Greenland 437
   ground 438–9
   hammerhead 439
   leopard 434, 438
   mackerel 440–1
   neckless carpet 434
   Pacific angel 434
   porbeagle 441
   Port Jackson 443
   salmon 440
   sand tiger 441
   shortfin mako 441
   smallfin gulper 436
   spined pygmy 436
   thresher 440–1
   whale 442–3
   zebra 443
shearwaters 250–1
sheep 198–9
   Barbary 193
   bighorn 199
   Dall's 199
   mountain 192–3
shells 349
shorebirds 282–3, 286–7
shrews
   elephant 78, 214–5
shrike, long-tailed 321
shrike-tit, crested 324
shrimp 515–16
   banded coral 516
sidewinder 391
sifaka
   diademed 108
   Verreaux's 106
silverfish 514–17
Siphonops annulatus 408
siren, dwarf 401, 402, 405
Sitta
   carolinesis 327
   europea 327
skate 445
   blue 446
   Port Davey 446
skimmer, black 285
skinks 376–7
   five-lined 343, 362, 379
   pink-tongued 377
   pygmy blue-tongue 378
skua 283
skulls 54–6, 58–61. 63–5, 120, 158, 203, 216, 290, 296

skunks
    Humboldt's hog-nosed 139
    Molina's hog-nosed 139
skylarks 318
sloth 81
    maned three-toed 92
    pale-throated three-toed 92
slow worm, European 362
slugs 495
snails 494–5
snakefly 37
snakes 343, 380–95
    African twig 386
    Arafura file 380, 386
    blotched pipe 380, 383
    blunt-headed tree 388
    colubrid 386–9
    coral 392–3
    Drummon Hay's earth 383
    Eastern ribbon 388
    elapid 390–3
    flowerpot 380, 382
    grass 387
    green vine 389
    king 387
    long-nosed tree 380
    mandarin rat 388
    pipe 382–3
    red cylinder 383
    red-bellied black 393
    sea 393
    South American coral pipe 383
    spotted harlequin 388
    spotted house 387
    sunbeam 380, 382
    thread 380, 382
    toad-eater 387
    white-bellied mangrove 386
snipe, Lathma's 284
snow leopard 148, 152
soapfish, sixline 480
solpugid 504, 512

Somateria
    mollissima 263
    spectabbilis 262
Somniosus microcephalus 437
Sooglossus sechellensis 420
Spalacopus cyanus 227
sparrows 236, 314, 328–30, 329
sparrowhawk 264–5
Speothos venaticus 127
Spermophilus columbianus 220
Sphecidae family 546
Sphecotheres viridis 332
Sphenodon sp. 361
Sphingidae family 544–5
spiders 503–11
    African signature 504, 510
    black widow 509
    bolas 508
    Brazilian wandering 509
    brown widow 509
    comb-footed 506
    crab 505, 506
    curved spiny 511
    funnel-web 508
    golden silk 511
    huntsman 504, 505, 509
    jumping 506, 508
    long-jawed orb weaver 511
    lynx 506
    marbled orb weaving 511
    mouse 508
    northern black widow 508
    orb-weaving 510–11
    primitive 506
    red widow 509
    redback 491, 508
    spiny orb weaver 511
    spitting 504, 508
    trapdoor 504, 508
    violin 508
    white lady 506
Spilocuscus sp. 88
Spilornis cheela 267

spinebill, Eastern 326
Spizaetus ornatus 269
Spiziapteryx circumcinctus 270
sponges 491, 492–3
Spongia officinalis 493
spookfish 445
spoonbill, roseate 257
sprat, Caspian 458
Squaliolus laticaudus 436
Squalus acanthias 436
Squatina californica 437
squid 495–7
    flying 496
    giant 496
squirrel glider 88
squirrels 25, 78, 216
    Abert's 219
    American red 220
    black giant 220
    Columbian ground 220
    delmarva fox 220
    Douglas' 219
    eastern fox 219
    Eurasian red 217, 221
    flying 221
    gray 219
Staphylinidae family 533
starfish 28, 490, 500
    crown-of-thorns 501
starlings 333
Steatornis carpensis 299
Stegosaurus 43, 47, 49, 62, 347
Stellula calliope 303
Stenella longirostris 175
Stenopus hispidus 516
sterlet 453
Sterna
    caspia 287
    paradisaea 287
Sternoptychidae family 469
Sterntherus minor minor 352
Stichopus chloronotus 501
stickleback 472

stingrays 444–7
  ocelated freshwater 446
  sixgill 446
Stomiidae family 468–9
Stomiiformes order 468
stork 343
  painted 236, 257
  saddle-billed 254
strawberry 15
Streptopelia turtur 288
Strix occidentalis 297
Struthiomimus 46, 56
sturgeon 448, 452
Sturnus vulgaris 333
Sturt's pea 18
Styrachosaurus 49, 64
sunbird, regal 327
sunbittern 279
Suricata suricatta 143
Sus
  barbatus 181
  scrofa 181
swallows 237, 318–19
swans 261, 263
swifts 301–3
  Alpine 303
  little 302
swordtail 474
Sylvia borin 325
Sylvilagus bachmani 230
Syncerus caffer 198
Synchiropus splendidus 474
Synodus variegatus 468
Syrmaticus reevesii 277

T

Tabanidae family 536
Tachybaptus ruficollis 249
Tachyglossus aculeatus 84
Tachymarptis melba 303

tahr, Himalayan 199
tails 54, 62–3, 125, 163, 363, 429, 438, 444
takahe 280
Talpa europaea 95
tamandua 92
tamarin
  cotton-top 105, 117
  midas 116
Tamias
  minimus 220
  striatus 219
Tamiasciurus
  douglasii 219
  hudsonicus 220
tanagers 328–9
Tangara fastuosa 328
tapeworms 491, 498
tapirs 178, 206–7
  Baird's 207
  Malayan 207
  mountain 207
  South American 207
Tapirus
  bairdii 207
  indicus 207
  pinchaque 207
  terrestris 207
tarantula 491, 504, 506
Tarentola mauritanica 373
tarpan 205
tarsier 104, 106
  Philippine 109
  spectral 108
Tarsius sp. 104, 106
  spectrum 108
  syrichta 109
Tasmanian devil 87
Tauraco leucotis 294
Taxidea taxus 138
teal, cinnamon 263
teeth 49, 56, 107, 122, 170, 355, 390, 435, 507, 509

temperate regions 22–3
temperature, regulation of 345
tenrec 95
  streaked 95
Tephritidae family 539
Terathopius escaudatus 269
Teratoscincus scincus 371
Teratornis merriami 241
termites 522–3
terns 282–3, 287
Terrapene carolina carolina 352
terror-bird 240
Tetracerus quadricornis 195
Tetraodon biocellatus 484
Tettigoniidae family 524
Thamnophis sauritus sauritis 388
Thelotornis capensis 386
Theropithecus gelada 112
theropods 56–9
thornbill, purple-backed 300
thorny devil 366
thrush, scaly 314, 322
Thymallus thymallus 466
tick 503, 512–13
tiger 79, 123, 148, 151, 581
Tiliqua sp 378
  scincoides 376
tinamou 236, 245
titi, masked 117
titmouse, plain 327
tits 314, 236–7
toadfish, splendid 470
toads 400, 410–21
  Asian horned 400, 410, 412
  Asiatic climbing 417
  crucifix 414
  Eastern narrow-mouthed 421
  European green 417
  Everett's Asian tree 416
  leopard 416
  natterjack 400
  oriental firebelly 400, 410, 413
  ornate horned 415

toads *continued*
  spadefoot 400, 410, 413
  Surinam 400, 410, 412
  yellow-bellied 412
Tockus sp. 309
tody, Cuban 307
Tolypeutes tricinctus 92
Tomistoma schlegelii 359
tongues 363
Torosaurus 65
tortoises 342, 348–51
  land 349–50
  saddleback 351
toucanet, emerald 313
toucans 236, 310–13, 580
  keel-billed 312
  toco 310
Trachichthys australis 477
Trachipterus sp. 477
Trachyphonus erythrocephalus 312
Trachypithecus obscurus 92
Tragelaphus eurycerus 194
tree dragon 367
tree fern 41
tree shrews 97
treeswift, gray-rumped 302
Tremarctos ornatus 130, 133
Triakis semifasciata 438
Triceratops 45, 46, 64
Trichechus
  inunguis 213
  manatus 213
  senegalensis 212
Trichoglossus haematodus 292
Trichuris sp. 498
triggerfish, clown 472, 485
trilobites 36–7, 39
Trimeresurus wagleri 394
Triturus marmoratus 406
Trogon
  violaceus 304
  viridis 305

trogons 304–5
  narina 304
  red-headed 305
  violaceous 236, 304
  white-tailed 305
Troödon 49
trout 448
  golden 466
  lake 466
  rainbow 466
Typhlonarke aysoni 446
trumpetfish, Caribbean 478
tuatara 342, 360–1, 583
tube worms 499
tuna 449
Tuojiangosaurus 63
Tupaia
  glis 97
  tana 97
turacos 294–5
Turdus migratorius 322
turkey 236, 275
turnstone 282, 283
Tursiops truncatus 175
turtles 31, 73, 342, 348–53, 583
  big-headed 350
  Eastern box 350
  Eastern spiny softshell 351
  loggerhead 348, 350
  Malayan snail-eating 350
  Pacific hawksbill 351
  painted 352
  softshell 349, 351
  spined 350
  twist-necked 342, 351
  yellow-spotted Amazon River 351
tuskfish, harlequin 482
tusks 178, 210
Tympanuchus cupido 276
Typhlonectes
  compressicauda 409
  natans 409

Tyrannosaurus 45, 46, 49, 57
Tyrannus
  tyrannus 317
  vociferans 316
Tyto alba 297

U

uakari 117
Uintatherium 83
Uncia uncia 152
ungulates 178–205
Uraniidae family 545
Urochordata class 493
Urocyon cinereoargenteus 129
Uroderma bilobatum 102
Ursavus 83
Ursus
  americanus 130–3
  arctos 132
  thibetanus 130, 136
urutu 394

V

Vanellus tricolor 284
Varanus
  giganteus 375
  gouldii 375
  komodoensis 374
  prasinus 374
  salvadorii 375
Varecia variegata rubra 108
Velociraptor 45, 58
Vespidae family 546, 548
Vicugna vucugna 184
vicuña 184
viperfish 469

vipers 390–1, 394–5
   rhinoceros 343, 380, 395
   Wagler's palm 394
vireos 329
vole, bank 225
Vombatus ursinus 88
Vormela peregusna 138
Vulpes macrotis 129
Vultur gryphus 272
vultures 264–5, 272–3, 580
   bearded 264
   palm-nut 272

W

waders 282–5
wagtails 318–19
waldrapp 257
wallabies 78, 90–1
   forest 90
   yellow-footed rock 91
walrus 122, 159, 161
wapiti 191
warblers 324–5, 329
   crested-sided 328
warthog 181
wasp 502, 518
waspfish, cockatoo 479
wasps 546–9
   European 548
   Ichneumon 548
   potter 546
   sand 546
water chevrotain 178, 190
waterfowl 260–3
waxwing
   Bohemian 320
   cedar 237
weasel 123
   African striped 141
   least 141

web of life 36–7
webs 505, 510–11
weevil 491
   giraffe 532
   Papua New Guinea 532
Weilandiella 41
weka 280
whales 162–5
   baleen 165–9
   beluga 170, 173
   blue 78–9, 162, 169, 582–3
   bowhead 162, 169
   false killer 173
   feeding 166–7
   fin 166–8
   gray 162, 168
   humpback 79, 163, 166–7, 169
   killer 163, 174–5
   minke 165, 167, 168
   northern bottlenose 173
   pilot 170
   pygmy right 162, 165, 168
   pygmy sperm 162, 173
   sei 163, 167
   southern right 165, 169
   sperm 162, 164, 170–3
   strap-toothed 162, 172
   tails 163
   toothed 164, 170–3
wheat 18
whipworms 498
wildebeest 196–7
   blue 194
willet 283
wings 73, 264, 285, 286
wobbegong, ornate 443
wolf
   gray 122, 124, 127
   maned 127
   red 127
wombat, common 88
woodchuck 219
woodlands 24–5

woodlark 318
woodpeckers 310–13
worm-lizards 343, 360, 362, 583
worms 491, 498–9
wormsnakes 382
wrens 322–3

X

Xanthomyza phrygia 326
Xantusia henshawi 377
Xenarthrans 92–3
Xenodon rabdocephalus 387
Xenopeltis unicolor 382
Xenopus gilli 413
Xiphophorus helleri 474

Y

yak 198

Z

Zaglossus bruijni 84
Zanclus cornutus 481
zebra 179, 202–5
Zenaida asiatica 288
Zonotrichia albicollis 328
Zoothera dauma 322
Zygaenidae family 545

Acknowledgments

Weldon Owen would like to thank the following people for their assistance in the production of this book: Philippa Findlay and Sarah Plant at Puddingburn Editorial Services, Andrew Davies at Creative Communications, Brendan Cotter, Helen Flint, Grace Newell and Guy Troughton.

The following illustrations are reproduced with permission from Magic Group, Czech Republic: (t=top, tl=top left, tc=top center, tr=top right, cl=center left, c=center, cr=center right, b=bottom, bl=bottom left, bc=bottom center, br=bottom right).

78tc, 79tr, 92bl, 94tr, 102tr, 106t,cl, 108tl,bl,br, 109t,tr,br, 111cl,cr, 112c,bl, 113tl,cr,bl,br, 114t,cl, 115tr,bl, 116t,cl, 117br, 138t,bl,br, 139tr,cl,cr, 140c,b, 141tl,tr,cl,cr,bl,bc,br, 159tc, 160tr,c,bl,br, 161tl,tr, 178tc, 180b, 181bl,br, 186b, 190cr, 201c, 205tc, 206tr,c,bl, 207tl,tr,c,bc, 212c,b, 213t,b, 215tl,tr,cl,br,bl, 216tl,tc, 217c,br, 223tl, 224cl,c,cr, 225c,bl,br, 226br, 227tl, 230tr,cl,c,cr,bl,br, 231tl,tr, 237tr, 249cl,br, 250cl, 270br, 274tr, 276c, 278tr,cl,cr, 281bl, 291br, 292cl,bl, 293br, 294bl,bc,br, 295br, 298tr,cl,c,br, 299tr,cl, 301tc, 302tr,br, 303cl, 304c, 305tl,tr,c, 307tl,cr, 308tr,cr, 311cl, 313tl,c,bl,br, 314t, 320cl, 321bl,br, 332br, 334cl,bl,br, 366bl, 368tr,cl,cr, 370b, 371b, 373b, 374c, 375t, 377br, 378cl, 382br, 383t,tr,bl, 392cl, 400tl,tc, 401tc,tr, 402tr, 404cl, 405c, 406tr, 407b, 408cl, 409cl,bl,br, 410tl,tr, 412tr, 413cl,c, 414t,cl,br, 415br, 416tr, 417tr,br, 426tc,br, 427tl,tc,tr,bl,bc, 432tr,c,bc, 448tl,tc,tr, 452tr,c, 453tc,c, 454tr,bc, 455bc, 458tr,cl,cr,bl,br, 459tc,c, 460tr,cl,bc,br, 461bc, 464tr,bc, 466tl,tr,c,bc, 467bc, 469tl, 470tr, 471bl,br, 476cl, 491tl, 493tl,tc,tr, 496cl, 500bl,br, 501tl, 514tl,tr,bl, 565cr, 566tc, 567cr, 568tr,c

Illustrators: Susanna Addario, Art Studio, Mike Atkinson/Garden Studio, Paul Bachem, Alistair Barnard, Priscilla Barret/Wildlife Art Ltd., Jane Beatson, Sally Beech, Andre Boos, Anne Bowman, Peter Bull, Martin Camm, Stuart Carter/Wildlife Art Ltd., D. Cole/Wildlife Art Ltd, Marjorie Crosby-Fairall, B. Croucher/Wildlife Art Ltd., Marc Dando/Wildlife Art Ltd., Peter David, Fiammetta Dogi, Sandra Doyle/Wildlife Art Ltd., Gerald Driessen, Simone End, Christer Erikson, Alan Ewart, Lloyd Foye, John Francis/Bernard Thornton Artists UK, Peg Gerrity, John Gittoes, Mike Golding, Mike Gorman, Ray Grinaway, Gino Hasler, Tim Hayward/ Bernard Thornton Artists UK, Dr. Stephen Hutchinson, Robert Hynes, Ian Jackson/Wildlife Art Ltd., Janet Jones, Roger Kent, David Kirshner, Frank Knight, Angela Lober, Frits Jan Maas, John Mac/FOLIO, David Mackay, Martin Macrae/FOLIO, Rob Mancini, Karel Mauer, Iain McKellar, James McKinnon, David Moore/Linden Artists, Robert Morton, Colin Newman/Bernard Communications, Ken Oliver/Wildlife Art Ltd., Erik van Ommen, Nicola Oram, Photodisc, Maurice Pledger, Tony Pyrzakowski, Oliver Rennert, John Richards, Edwina Riddell, Steve Roberts/Wildlife Art Ltd., Barbara Rodanska, Trevor Ruth, Claudia Saraceni, R.T. Sauey, Michael Saunders, Peter Schouten, Peter Scott/Wildlife Art Ltd., Rod Scott, Chris Shields/Wildlife Art Ltd., Ray Sim, Marco Sparaciari, Chris Stead, Kevin Stead, Mark Stewart, Stockbyte, Roger Swainston, Steve Trevaskis, Thomas Trojer, Chris Turnbull/Wildlife Art Ltd., Glen Vause, Genevieve Wallace, Trevor Weekes, Rod Westblade, Ann Winterbotham.